THE AI EDGE

"The AI Edge *isn't like most AI books. It puts people back at the center of the story. Rather than hype or fear, this book offers practical clarity on how AI becomes a competitive advantage when paired with human judgment. Drawing on insights from dozens of experts across industries, this book delivers a timely, grounded guide for leaders who want to move faster without losing what makes them human.*"

—CARL CARPENTER, Senior AI Strategy Consultant

"*This book is an accessible and deeply thoughtful roadmap for anyone trying to make sense of AI's role in learning, work, and everyday life. The analogies are powerful, the message is empowering, and the diversity of expert voices makes the insights feel real and actionable. The AI Edge is not about becoming technical. It's about becoming adaptable.*"

—CINDY BOWEN, Adjunct Professor of Data & AI

"*Unlike many AI books written from a single vantage point,* The AI Edge *succeeds by bringing together practitioners who are actually using AI in the real world. The result is a multifaceted, honest, and surprisingly practical guide to how AI is reshaping industries. This is a must-read for anyone who wants to stay relevant as technology accelerates.*"

—JORDAN CYSZYNSKI, CEO of OnsetAI

"AI can feel overwhelming, especially for entrepreneurs and small business owners. The AI Edge cuts through the noise and shows how accessible these tools truly are. Each chapter stands on its own, offering concrete ideas that can be applied immediately. This book doesn't just explain AI, it helps you use it to reclaim time, focus, and momentum."

—JARVIS LEVERSON, Founder of Accelerated Growth Club

"What makes The AI Edge stand out is its balanced voice. It neither glorifies AI nor fears it. Instead, it frames AI as a tool that amplifies human capability when used thoughtfully. By emphasizing judgment and creativity, this book offers a refreshing perspective on how humans and machines can evolve together."

—KATTIE SYLVESTER, AI Researcher

"Clear, engaging, and reassuring, The AI Edge feels like a conversation with trusted guides rather than a technical manual. The personal stories and practical examples make complex ideas approachable. Whether you're new to AI or already experimenting, this book helps you see how to sharpen your edge without losing control of your life."

—SAMANTHA KOCH, Keynote Speaker on AI

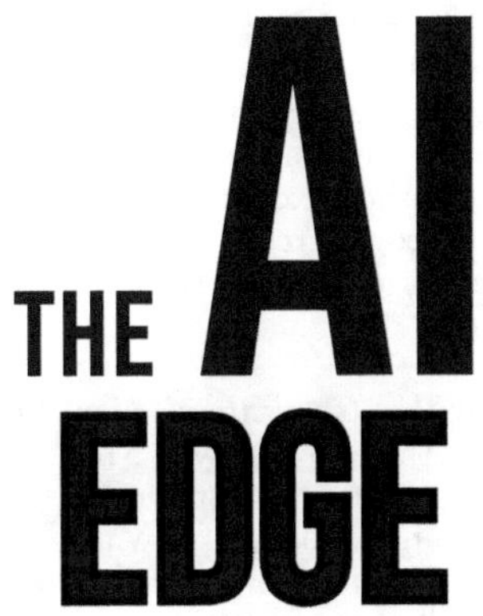

THE AI EDGE

How to Thrive Within Civilization's Next Big Disruption

Authored by:
Erik Seversen, Anupam Agarwal, MD, MPH, Hendo AI,
Pankhuri Bansal, Jason Bowden, Shang-Wei Chao, Christiano
Gerhart Dittrich, Steven G. Edwards, Russell Goldman, Robin
Goudappel, Sowmiya Narayanan Govindaraj, Jonathan Fraile
Gutierrez, Anat Heilper, Penny Hopkinson, Sara Jetta, Olga
Martynov, Agnieszka Mietz-Blijleven, PhD, Tevan Millette,
Gerald Pallor, Russell Ramesh, Annette Raynor, Doug Razzano,
Anthony Rizk, PhD, Christian Schenk, PhD, Amit Shivpuja,
Shelly Swanback, Sakina Syed, Leszek Tasiemski, James
Taylor, Megan Thomas, Andrew J. Turner, Aasim Waheed,
Daniel Wielander, Irene Yang, Joe Zhou

THIN LEAF PRESS | LOS ANGELES

Library of Congress Cataloging-in-Publication Data
Names: Seversen, Erik, Author, et al.
Title: *The AI Edge: How to Thrive Within Civilization's Next Big Disruption*
LCCN: 2026902993

ISBN 978-1-968318-50-5 (hardcover) | 978-1-968318-49-9 (paperback)
ISBN 978-1-968318-48-2 (eBook) | 978-1-968318-51-2 (audiobook)

Artificial Intelligence, Science & Technology, Business, Professional Development
Cover Design: 100 Covers
Interior Design: Dindo Sanguenza
Editor: Nancy Pile
Thin Leaf Press
Los Angeles

Thank you for reading this book. There is information found within the following pages that can greatly benefit your life, but don't stop there. Make sure you get the most you can from this book and reach out directly to the expert-authors who want to help give you an edge by using AI, to thrive within civilization's next big disruption, and to manifest success in your life. Contact information for each author is found at the end of their respective chapter.

To those at the cutting edge of new technologies that are
transforming the world in which we live.

CONTENTS

INTRODUCTION

By Erik Seversen
Author of *Ordinary to Extraordinary*
Los Angeles, California

*Every person, every company, every industry will be transformed
by AI. Those who learn how to use it will move faster
than those who don't.*
—Satya Nadella, CEO of Microsoft

When I was in elementary school, I remember struggling with my multiplication tables. One day, my teacher gave me a set of flashcards, and this was exactly the edge I needed to pass the math quizzes. Later in junior high, I was able to use a calculator to help with math and eventually a scientific calculator to give me an edge with my algebra classes. Using these tools made it much easier to complete the tasks in front of me, and once the math got to a certain level, I couldn't possibly keep up without these tools.

During high school, I played ice hockey. Every week, before our hockey games, I'd have to get my skates sharpened. After tons of laps for cardio, drills for hockey skills, and scrimmages during practice, my skates would become quite dull. Sharpening them gave me both a literal and metaphorical edge when it was gametime. My ability to skate at my best required that the blades of my skates were as sharp as they could be. With dull skates, I couldn't possibly keep up with the other players.

Now there is a relatively new tool which I use frequently that is allowing me to perform at my best levels. This tool is AI, or maybe more appropriately, these tools are AI. There are so many different types of AI, it is important to use the right AI tool for the right task. I strongly believe that just like the days of early calculators, basic word processing machines, and other inventions, people who use AI effectively are going to have an edge over those who don't.

What makes the current shift in AI different from past technological shifts is not simply the power of the tools, but their accessibility. AI is no longer confined to research labs, massive corporations, or elite technical teams. Today, a student, a small business owner, a writer, a manager, or an entrepreneur can access capabilities that were unthinkable just a decade ago. Tasks that once required teams of specialists can now be supported, or accelerated, by AI systems that learn, adapt, and assist in real time. The AI edge is no longer reserved for the few; it is available to anyone willing to learn how to use it.

This does not mean AI replaces human creativity or judgment. Just as a calculator does not understand mathematics and sharpened skates do not make someone a great hockey player, AI does not succeed on its own. The advantage comes from the human who knows when to use the tool, how to use it well, and how to integrate it into their thinking and decision-making. AI amplifies ability. It reduces friction and frees time. But most importantly, it allows people to focus on higher-level work, like strategy, creativity, problem-solving, and connection, rather than getting stuck in repetitive or limiting tasks.

The danger, however, lies in ignoring this shift. History shows that those who resist new tools out of fear, skepticism, or complacency are often left behind, not because they lack talent, but because they refuse the advantage available to them. This book is about recognizing that civilization is once again undergoing a major transformation and that thriving in times of disruption requires adaptability and the willingness to learn. As an educator, my primary purpose in creating this book is to help people gain an advantage, an edge, in both their work and their lives by intelligently using AI when it makes sense.

Since I'm not an AI expert, I didn't try to write this book about AI by myself. Instead, this book brings together insights from 34 experts working across diverse fields including technology, business, education, creativity, healthcare, research, and more. Each contributor approaches AI from a different angle, showing how specific types of AI can be applied to real-world challenges and opportunities. Together, these perspectives form a clear and grounded picture of how AI is already reshaping the way we work, think, and compete and how individuals can position themselves to benefit rather than fall behind.

The co-authors of this book come from all over the USA, Canada, the United Kingdom, the Netherlands, Germany, Poland, Austria, Brazil, the United Arab Emirates, Lebanon, Taiwan, and Australia. These authors have been featured on media outlets including NBC, CBS, FOX, and more. They are professionals who are university professors, business owners, consultants, product managers, engineers, software developers, cybersecurity leaders, healthcare experts, advisors, TEDx and keynote speakers, brand strategists, researchers, platform architects, senior data analysts, IT directors, CTOs, robotics experts, compliance officers, and more. The one thing these individuals have in common is that they all have something to share about artificial intelligence, and these ideas are available to you now.

Although this book is organized around the united theme of using AI as your edge to gain advantage within civilization's next big disruption, each of the chapters is totally stand-alone. The chapters in the book can be read in any order. I encourage you to look through the table of contents and begin wherever you want. However, I urge you to read all the chapters because, as a whole, they provide a great array of perspectives. Each is valuable in helping you understand how AI is being used in many areas of human industry.

It is my hope that you discover something in this book that helps you navigate civilization's next big disruption as humans and machines become more interdependent and AI technology continues to insert itself into many aspects of our everyday lives. It is my hope that you embrace AI as you create your AI edge within the global transformations that are occurring.

Whether you are just beginning your journey with AI or already experimenting with its possibilities, this book aims to help you develop an AI mindset that sees these tools as partners, not threats. By understanding how AI is being used in different industries, it is my hope that you gain the edge you need to be the best version of yourself at work and in life. Whether this version is a pinnacle of productivity or the freedom to slow down and spend more time with family and friends, this book is an invitation to sharpen your edge and step confidently into the new era of humans and machines working together.

About the Author

Erik Seversen is on a mission to inspire people. He holds a master's degree in anthropology and is a certified practitioner of neuro-linguistic programming. Erik draws from his years of teaching at the university level and years of real-life experience in business to motivate people to take action, creating extreme success in business and in life.

Erik is a TEDx and keynote speaker who has reached over one million people through his public speaking and live courses. He has visited 99 countries and all 50 states in the USA and has climbed the highest mountains on four continents, 15 countries, and 18 states. Erik has published 19 bestselling books on the topics of mindset, success, and peak performance, and he has helped over 400 people become authors. He is a full-time writer, book consultant, and speaker, and he lives by the idea that success is available to everyone—that living an extraordinary life is a choice.

Erik lives in Los Angeles with his wife and has two boys currently studying at university.

Contact Erik for interviews, speaking, or book publishing consultation.

Email: Erik@ErikSeversen.com

Website: www.ErikSeversen.com

LinkedIn: https://www.linkedin.com/in/erikseversen/

THE AI EDGE IN MEDICINE: INTEGRATING ARTIFICIAL INTELLIGENCE INTO CLINICAL CARE AND RESEARCH

By Anupam Agarwal, MD, MPH
Physician Executive in Biopharma & AI Strategy
San Francisco, California

The future of AI needs to be human-centered. It must be guided by human wisdom, human values, and human empathy.

—Fei-Fei Li, 2018 Stanford HAI Launch Keynote

A Turning Point in Modern Medicine

Modern medicine stands at a decisive inflection point. For decades, clinicians and researchers have imagined technologies capable of analyzing massive datasets, connecting patterns across imaging, genomics, and clinical notes, or identifying weak signals that humans naturally overlook. Today, these capabilities are not theoretical; they are entering daily practice. Hospitals deploy predictive models to warn clinicians of deterioration. Radiologists rely on AI-driven triage systems to identify and flag urgent findings within minutes. Research groups use deep learning algorithms to recognize complex molecular interactions or to identify signals within large pharmacovigilance databases.

The phrase "AI edge" describes the advantage physicians and scientists gain when artificial intelligence is integrated responsibly into clinical and research environments. The advantage does not arise from replacing clinicians, nor from delegating judgment to machines. Instead, it emerges when AI augments human reasoning, capturing detail, reducing noise, and accelerating cognition in ways that support, rather than supplant, clinical expertise.

This chapter explores how that advantage develops, where it is already visible, and why its success depends on rigorous evidence, transparent oversight, and a realistic understanding of AI's limitations. Each section combines high-level conceptual explanation with concrete examples drawn from peer-reviewed literature, regulatory decisions, or widely documented real-world implementations. The goal is clarity rather than hype, and a practical view of the future rather than a speculative one.

What AI Actually Does in Medicine

Artificial intelligence is not a single technology. It is a family of analytical approaches that differ in structure, purpose, and risk. Understanding these foundations is essential for anyone practicing medicine with AI-driven tools.

Machine Learning (ML)

ML algorithms detect patterns in numeric, textual, or categorical data. In medicine, ML underlies tools that estimate readmission likelihood and predict postoperative complications or identify patients at risk for serious clinical worsening. These systems excel when large samples and consistent patterns exist, but are sensitive to biased input data or poorly calibrated thresholds.

Deep Learning (DL)

Deep learning models use neural networks with many layers to extract complex features from images, waveforms, and text. DL has driven major advances in radiology, dermatology, ophthalmology, and pathology. Tools such as convolutional neural networks (CNNs) have demonstrated performance comparable to specialists in specific tasks, for example, detecting diabetic retinopathy, classifying chest radiographs, or identifying melanoma in dermoscopic images.

Large Language Models (LLMs)

LLMs interpret unstructured clinical text, draft documentation, summarize literature, and generate coherent responses by training on extensive language datasets. Their strength lies in language manipulation; they do not inherently "understand" physiology or mechanisms of disease. Responsible clinical use, therefore, requires clearly defined boundaries, human review, and protective workflows.

Agentic Systems

A newer generation of tools can initiate multi-step processes—such as drafting entire clinical notes, preparing prior-authorization documents, or screening safety cases—under supervision. These systems can meaningfully reduce administrative load but require strong oversight to avoid cascading errors.

The Non-Negotiable Foundations of Trustworthy AI

Regardless of model type, safe implementation depends on:

- Data quality and representativeness
- Bias detection and mitigation
- Transparent training and validation pipelines
- Calibration and local testing before deployment
- Continuous monitoring and version control
- Clear human accountability at every decision point

These elements determine whether an AI system strengthens clinical care or inadvertently introduces risk.

AI AT THE POINT OF CARE

Diagnostic Support

Diagnostic AI has matured more rapidly in imaging-heavy specialties, where structured data and large labeled datasets are available.

Radiology

Several FDA-cleared systems exemplify responsible deployment. IDx-DR, approved for diabetic retinopathy screening, demonstrated its ability to identify referable retinopathy with high sensitivity and specificity in a large multicenter study. Viz.ai, cleared for automated stroke detection and workflow optimization, reduces time to notification for large vessel occlusion and accelerates neurology team mobilization.

Research has also shown that deep-learning models can classify chest radiographs, detect pulmonary nodules, identify pneumothorax, and analyze breast imaging with performance that approaches or occasionally matches experienced radiologists. Yet, real-world implementation consistently demonstrates that models may perform differently across sites due to variation in imaging equipment, population demographics, or documentation practices. Therefore, every model, regardless of publication quality, must be tested locally.

Cardiology

Cardiology has seen remarkable advances as well. Several peer-reviewed studies have shown deep-learning systems can identify left ventricular dysfunction directly from standard ECG waveforms, sometimes before overt clinical symptoms appear. Wearable devices, including those with validated atrial fibrillation detection algorithms, have demonstrated real-world detection performance in large prospective cohorts.

Dermatology and Pathology

AI-assisted lesion classifiers can aid in identifying suspicious skin lesions, although performance depends heavily on skin-tone diversity in training data. Digital pathology systems support tumor grading, mitotic count identification, and margin assessment, easing workload in high-volume centers.

Across disciplines, the AI edge emerges when clinicians use these tools as structured inputs, not replacements, for diagnostic reasoning.

Predictive Analytics and Clinical Decision Support

Predictive models identify subtle but meaningful indicators of future risk.

Early Warning Systems

Machine-learning approaches have been deployed to predict sepsis, ICU transfer, or postoperative deterioration. Some real-world implementations have documented earlier detection compared with rule-based systems like MEWS or SIRS criteria. However, large-scale analyses have shown that the performance of some commercial sepsis models declines significantly when deployed systemwide. These findings reinforce the principle that AI is not inherently reliable; it must be continuously validated and monitored.

Cardiovascular Risk and Acute Events

Predictive analytics help estimate the risk of heart failure hospitalization, arrhythmias, and major adverse cardiac events. These models rely on rich combinations of EHR data, imaging, biomarkers, and longitudinal vitals. When calibrated properly, they support earlier intervention and more tailored follow-up.

Resource Allocation and Workflow Optimization

Hospitals increasingly use predictive models to anticipate imaging demand, triage emergency department patients, or forecast bed availability. These systems help administrators allocate staff, manage bottlenecks, and maintain throughput during peak hours.

Streamlining Clinical Operations

Administrative burden is a well-documented contributor to physician burnout. AI tools designed to reduce such burden can meaningfully improve clinical workflow.

Documentation

Early studies indicate that LLM-based assistants can reduce after-hours charting by drafting visit summaries, generating structured problem lists, and preparing preliminary documentation for clinician review. These systems do not replace the clinician's narrative or judgment, but they can eliminate repetitive clerical tasks.

Patient Communication and Triage

Models that categorize incoming messages or triage requests based on urgency help reduce response delays. Automated scheduling assistants can coordinate follow-up visits, route tasks appropriately, and reduce time spent on low-complexity logistics.

Operational Efficiency

AI-supported bed management, staffing prediction, and capacity forecasting tools improve hospital flow and allow clinical teams to focus on patient care rather than manual coordination.

These operational applications rarely capture headlines, yet they may yield some of the most immediate benefits in real clinical settings.

AI IN DRUG DEVELOPMENT AND CLINICAL RESEARCH

Target Identification and Preclinical Discovery

Drug discovery traditionally requires navigating immense biological and chemical complexity. AI helps by accelerating several stages of early research.

Protein Structure and Mechanistic Insight

Deep-learning breakthroughs in protein structure prediction, most notably AlphaFold's performance reported in peer-reviewed evaluations, have influenced basic research by offering high-quality structural estimations for proteins lacking crystallographic data. While not a replacement for experimental work, these models allow teams to generate hypotheses more rapidly.

Generative Chemistry

Machine-learning models can explore large chemical spaces and generate new molecular structures predicted to have desirable properties. Several clinical-stage programs have originated from AI-assisted molecular design pipelines, though long-term outcomes remain under evaluation.

Biomarker Discovery

Algorithms trained on multi-omics datasets can identify potential biomarkers or molecular signatures for disease subtypes, informing precision medicine strategies and patient stratification.

Trial Design, Execution, and Analysis

AI influences multiple aspects of the clinical trial lifecycle.

Synthetic Control Arms

In rare diseases or situations where recruitment is challenging, synthetic control arms built from real-world datasets offer a viable alternative to traditional controls. Regulators have accepted these designs in specific, well-documented contexts when supported by transparent statistical methods and rigorous data curation.

Adaptive Trial Design

Machine-learning models can assist with interim assessment, enabling adaptive randomization or early stopping decisions under pre-specified conditions. Although the statistical rigor still rests with human biostatisticians, AI enhances efficiency by processing complex, multi-dimensional data in real time.

Site Selection and Enrollment Optimization

Historical EHR and claims data can help identify trial sites with high concentrations of eligible patients or forecast enrollment rates. AI-supported feasibility analyses reduce trial delays and improve resource allocation.

Real-World Evidence Generation

Following approval, AI tools assist in analyzing registry data, EHR records, and post-market surveillance databases, supporting regulatory submissions and safety monitoring programs.

Pharmacovigilance and Safety Intelligence

Pharmacovigilance relies on timely detection and accurate interpretation of adverse events across large datasets. AI is well suited to this domain.

Signal Detection

Machine-learning approaches can reduce noise in spontaneous reporting systems, improving sensitivity to emerging patterns. These systems highlight combinations of events or temporal trends that might otherwise go unnoticed.

Case Triage and Narrative Processing

Natural-language processing tools classify seriousness, extract relevant details from clinical narratives, and prioritize cases for human review. Several companies and research groups have published results showing meaningful reductions in manual workload.

Duplicate Detection and Literature Surveillance

AI systems can identify likely duplicates across reporting systems, improving database quality. Automated literature surveillance tools scan journals and preprints, flagging potentially relevant safety publications for expert assessment.

Regulators have acknowledged these applications, emphasizing that AI may support, but cannot replace, human pharmacovigilance judgment.

INTEGRATION CHALLENGES: WHAT SLOWS ADOPTION

Data Fragmentation and Interoperability

The limiting factor in AI performance is frequently the underlying data. EHR systems vary dramatically in structure, coding practices, and completeness. Imaging repositories may lack standardized metadata. Claims data capture reimbursement details but often omit key clinical variables, such as smoking status and family history. These structural limitations can restrict even the strongest models.

Efforts such as FHIR, standardized vocabularies, and common data models improve interoperability but require consistent implementation across institutions.

Ethical and Legal Considerations

AI raises important ethical questions that influence trust and adoption.

Bias and Fairness

Models trained on populations lacking diversity may perform less accurately on underrepresented groups. This concern is not theoretical.

Multiple published examples demonstrate disparities in predictive model performance across race and ethnicity.

Privacy and Data Protection

De-identified data can sometimes be re-identified when cross-referenced with other datasets. Regulations such as HIPAA, GDPR, and state-level privacy laws shape how data may be collected, stored, and used.

Accountability

Even when algorithms contribute to decision-making, clinicians are ultimately responsible for patient care. Institutions must clearly define governance, responsibility, and escalation pathways to avoid ambiguity.

Human Factors and Clinical Culture

AI succeeds or fails based on whether it fits the reality of clinical practice.

Systems that introduce friction—extra screens, unclear alerts, or confusing outputs—are rapidly abandoned. Conversely, tools that integrate naturally into existing workflows can quietly deliver substantial benefits.

Clinician trust grows when systems:

- provide transparent reasoning or explainability
- demonstrate consistent real-world performance
- incorporate clinician feedback
- avoid overwhelming users with false alarms

These cultural factors are as critical as the underlying algorithms.

Regulatory and Governance Frameworks

Regulatory clarity has expanded in recent years.

FDA and International Regulators

The FDA regulates diagnostic and predictive AI systems as software as a medical device (SaMD). Clearance pathways require demonstration of safety, effectiveness, and risk mitigation. The agency has also proposed frameworks for adaptive algorithms that may update over time, emphasizing pre-specified change-control plans and post-market monitoring.

The EMA and other regulatory bodies similarly emphasize transparency, reproducibility, and rigorous evidence. Although terminology varies, global regulators converge on several guiding principles:

- clear description of training datasets
- robust validation using representative populations
- monitoring for drift and bias
- traceable documentation of model updates

Institutional Governance

Hospitals and health systems have begun creating AI governance boards. These groups typically include clinicians, data scientists, ethicists, legal experts, and quality officers. Their responsibilities include:

- reviewing proposed models
- evaluating evidence before deployment

- defining acceptable performance thresholds
- establishing ongoing monitoring programs
- determining when updates or deactivations are required

Governance frameworks allow clinicians to adopt AI confidently, knowing oversight mechanisms exist.

The New Clinical Skill Set: Practicing Medicine with AI

Integrating AI into daily practice expands the clinician's role. Clinicians must develop literacy in:

- understanding model calibration and confidence measures
- recognizing when outputs are unreliable
- distinguishing between correlation and causal inference
- identifying situations where AI may amplify bias
- interpreting structured outputs rather than accepting them at face value

Medical schools are beginning to incorporate coursework on informatics, data interpretation, and AI ethics. Continuing professional development programs increasingly teach model evaluation, human-AI collaboration strategies, and practical implementation principles.

AI does not diminish clinical reasoning. It elevates the demands placed upon it.

Building AI-Enabled Health Systems

Health systems that succeed with AI share several structural characteristics.

Robust Data Pipelines

They invest in clean, well-curated datasets with consistent coding, high-quality metadata, and secure storage. These systems understand that strong data infrastructure is the foundation on which every model relies.

Seamless Workflow Integration

They embed AI into the clinician's existing environment, within the EHR, imaging viewer, or workflow tool, so that interacting with AI requires no additional steps.

Continuous Monitoring

They track drift, monitor false-positive rates, and evaluate performance across diverse populations. They require formal review before deploying new model versions.

Organizational Culture

They encourage transparency, clinician feedback, and shared ownership of safety. They acknowledge that no model is perfect, and they build systems that can detect errors quickly. These characteristics are more predictive of success than the sophistication of any individual algorithm.

Vision: A Human-Centered, AI-Augmented Future

AI will become increasingly embedded in medicine, from diagnosis to workflow optimization to safety assessment. Yet the greatest value will come from amplifying, not replacing, human clinical judgment.

Clinicians will spend less time on clerical burdens and more time explaining diagnoses, aligning goals of care, guiding families, and integrating complex biological information. Researchers will navigate molecular and clinical datasets with unprecedented efficiency. Pharmacovigilance teams will detect safety signals earlier and with greater precision. Patients will benefit from faster diagnosis, more coordinated care, and reduced delays.

The future of AI in medicine is not about automation; it is about augmentation. The AI edge emerges when clinicians maintain responsibility, regulators provide clear oversight, and institutions build the infrastructure needed for safe, transparent deployment.

Conclusion

Artificial intelligence has advanced from theoretical concept to practical clinical instrument. It strengthens diagnosis, improves prediction, accelerates discovery, and supports safety monitoring when deployed with rigor. The AI edge lies not in the algorithms alone but in the disciplined integration of validated models, governance frameworks, transparent data practices, and human clinical reasoning.

Medicine will continue evolving, but the values that define it—judgment, compassion, and responsibility—remain firmly in human hands. AI's role is to enhance those values, not compete with them.

About the Author

Anupam Agarwal, MD, MPH, is a physician–executive, cardiologist, and leading advisor to biotech founders and investors at the intersection of AI and medicine. With more than 20 years of experience in global biopharma and advanced-therapy development, he has guided clinical, safety, and regulatory strategy across cardiovascular disease, rare disorders, neuroscience, and gene therapy. He has served as senior vice president and global head of safety and pharmacovigilance and now supports emerging and growth-stage companies seeking to integrate AI-driven insights into drug development and safety science. A Harvard alumnus, he also contributed to *The AI Revolution* and *The AI Edge*.

ANUPAM AGARWAL, MD, MPH

Email: anupam_agarwal@post.harvard.edu

CREATIVITY UNBOUND: HOW AI UNLOCKS HIDDEN CREATIVE GENIUS

By Hendo AI
Award-Winning Creative Director & AI Strategist
London, England, United Kingdom

Inspiration exists, but it has to find you working.
—Pablo Picasso

The Night Everything Changed

For more than 20 years, I worked in the traditional creative industry. I understood how brands were built, how campaigns were shaped, and how ideas moved through organisations. The process was familiar, respected, and guarded. Yet it was also constrained by layers of approvals, resource limitations, and long timelines that slowed momentum and constrained innovation. Then, in November 2022, everything shifted.

Late one night while staying at a friend's house, after everyone had gone to bed, I opened my laptop and experimented with something I had only recently heard about: ChatGPT. Out of curiosity, I asked it to write a creative strategy for an emerging brand. Within seconds, it produced a structured response that would normally take me days of work. In that moment, I realized that creativity was about to enter a new era. The output wasn't the revelation. The speed was. And speed changes who wins. And I knew immediately that I had a decision to make: Remain in the safety of the old world, or step into the acceleration of the new. I chose acceleration.

A New Creative Frontier

In the days that followed, the possibilities of what I had witnessed became impossible to ignore. It wasn't just the novelty of AI's output that struck me; it was the speed, access, and removal of barriers that had shaped the traditional creative process. Shortly after, I began experimenting with Midjourney, generating surreal visual scenes in minutes that once would have required teams, budgets, equipment, and long schedules. Creativity was no longer bound to traditional structures, resources, or approvals. Ideas could be explored instantly, tested rapidly, and iterated upon continually.

While conventional organisations continued debating concepts in boardrooms, early adopters were publishing, learning, improving, and evolving at remarkable pace. The gap between traditional workflows and AI-enabled creation widened quickly. For leaders who built their careers in a more traditional landscape, this shift can feel disruptive. But embracing AI doesn't diminish the value of prior experience, it extends it. The same instincts, judgment, and discipline that shaped past success become even more powerful when combined with the capabilities of AI. This is not about replacing what got us here, but about ensuring it continues to lead us forward.

Breaking the Bottlenecks

Before AI, creativity, whether developing campaigns, shaping strategy, or designing new experiences, was limited by logistics. Budgets, scheduling, approvals, revisions, and departmental dependencies slowed even the most compelling ideas. A single initiative could take months to reach an audience. That pace made sense when the only way forward was slow, careful execution.

Today, that model is no longer competitive. AI collapses bottlenecks across ideation, production, communication, and deployment. Brands using AI are testing concepts in real time, learning from immediate feedback loops, and scaling successful ideas quickly. One of my first viral experiences came from an AI-generated image of the cast of *The Matrix* having a secret rave in a basement. It took only minutes to create, purely as an experiment, yet it travelled across platforms, was copied and remixed globally, and accumulated millions of impressions. It proved something essential: A playful idea, executed instantly, can outperform a meticulously managed traditional campaign.

"Traditional creation scales slowly. AI-enabled creativity, strategic and commercial, scales immediately".

Don't post to go viral. Post to become the source code.
March 2025

AI is not only accelerating creativity. It is accelerating entrepreneurship. Individuals are now using AI to identify market opportunities, design applications, and generate the content required to promote them, all within days rather than months. People are discovering real-world problems, creating AI-enabled solutions, and then using AI again to drive awareness and adoption at scale.

The speed at which a new idea can be validated, launched, and refined has changed completely. This isn't simply faster creativity;

it is accelerated innovation. Beyond content and creative output, AI is rapidly reshaping core business systems. I'm seeing organisations use AI to streamline supply chains, improve forecasting accuracy, automate customer service, accelerate R&D cycles, and simulate market outcomes before decisions are made. Teams are developing prototypes overnight, stress-testing ideas in hours, and reallocating resources based on real-time insight. What once required specialist departments, long meetings, and a significant budget now begins with a prompt, a model, and a leader willing to run the experiment. This is no longer about creativity in the artistic sense. It's creativity applied to operations, growth, and organisational advantage.

What AI Revealed

The deeper I went into this technology, the more I realised that AI was not replacing creativity. It was expanding it. For years, I believed creativity was centred primarily around the spark of the idea itself. AI revealed something far broader. Creativity is the entire system: generating the idea, testing its validity, refining it through feedback, optimising its positioning, and accelerating its execution. With AI, I could explore 20 variations of a concept instead of two. I could challenge assumptions, test hypotheses, and simulate strategies that once would have required extensive planning and resourcing.

Creativity became more ambitious, and importantly, more strategic. The brands adopting AI are not simply producing more content. They are improving decision-making, compressing innovation cycles, and uncovering new possibilities. AI does not diminish experience. It multiplies it. We are entering a moment where small experiments can produce outsized results. I have already seen teams reduce production cycles by more than half, increase output, and validate ideas in a fraction of the time they once required. When creativity is no longer bottlenecked by capacity, it becomes a vehicle for growth, not just expression.

Rethinking Creativity

Over time, I came to understand that the primary resistance to AI is not technological but psychological. Many experienced professionals hesitate because integrating AI challenges their sense of mastery. When an individual has developed expertise over decades, the idea that new tools could perform aspects of their craft instantly can be unsettling. But embracing AI does not erase experience. It elevates it. I have watched businesses of every size, regardless of sector, transform through AI adoption. They communicate more effectively, test strategically, deploy faster, and learn with greater clarity. Meanwhile, teams adhering rigidly to legacy processes are already struggling to maintain momentum and relevance. Creativity, in its truest form, is the ability to reimagine and improve. Teams that embrace AI are not only more adaptive and innovative. They become magnets for ambitious talent seeking to shape the future, rather than survive it.

Where We Go from Here

We are living through the fastest creative and strategic transformation in history. The tools we use today will look primitive compared to what is coming, and we are already seeing dramatic shifts in workflows, efficiencies, and market expectations. Early adopters of AI are learning at speed, leveraging insights rapidly, and compounding advantages across their organisations. If the last decade rewarded efficiency, the next will reward adaptability.

We are moving into a landscape where businesses will evolve continuously, not annually. Roadmaps will compress, teams will deploy faster, and leaders will need to make decisions with more experimentation and less certainty. The organisations that thrive won't be the ones with the most resources, but the ones willing to learn, adjust, and build with creative momentum. AI gives us the ability to do that at scale, not with chaos, but with clarity.

Meanwhile, businesses delaying involvement are losing momentum quietly but significantly. The advantage gained through early adoption compounds rapidly. Each experiment builds confidence, each iteration builds skill, and each insight improves the next. Over time, these gains create a separation that becomes increasingly difficult to close. What begins as a learning phase quickly becomes a competitive advantage and eventually, a strategic moat.

This moment is not about discarding the strengths of the past. It is about evolving them into the future. The opportunity now is to explore what is possible, expand what creativity includes, and build strategically for the world that is emerging. The future is accelerating, and the window for hesitation is narrowing. The next step is yours to take. Because the threat isn't AI itself, it's the competitor who adopts it before you do. The real question for leaders is no longer whether AI matters. It's how long you can afford to wait before acting. Six months of inertia could become 18 months of catch-up. A competitor who experiments today may outpace you without ever hiring a larger team.

The shift is already underway, and it's not slowing down for anyone. If your organisation has experience, brand equity, and momentum behind it, imagine what becomes possible when velocity is added to that foundation. The potential is enormous, but it rewards movement, not observation.

We Are at the Beginning

For organisations unsure where to begin, the first step isn't transformation. It's exploration. Start small. Run pilot tests. Assign internal champions. Create a protected environment to explore the tools without pressure. Curiosity alone is enough to begin building capability, and momentum grows naturally through learning.

Most organisations don't need a full transformation to begin, they need a foothold. A 30-day pilot, a single workflow experiment, or a small creative sprint can reveal more than a year of theoretical discussion. Once momentum begins, confidence follows. The most important realisation for leaders today is that AI is not a distant horizon; it is already reshaping the present. Companies moving boldly are gaining advantage, while those waiting for certainty are losing time they cannot recover. This moment is not about discarding the strengths of the past. It is about evolving them into the future.

About the Author

Hendo AI is an award-winning creative director, brand strategist, and AI filmmaker with over 20 years of experience leading creative and strategic work across technology, entertainment, lifestyle, and professional services. His work focuses on building strategic leverage through AI, integrating artificial intelligence into creativity, systems, and decision-making to accelerate innovation and competitive advantage.

Today, he works with leaders, teams, and organisations to apply AI in practical, high-impact ways across storytelling, content systems, and business transformation. Through consulting, advisory work, workshops, and collaboration, he helps organisations move from experimentation to confident execution in an accelerating AI-driven landscape.

Website: www.hendoai.com

CHAPTER 3

THE ERA OF COGNITIVE PARTNERSHIP: HOW HUMANS AND AI THINK BETTER TOGETHER

By Pankhuri Bansal
AI/Blockchain Expert, United Nations, Founder, Speaker
London, England, United Kingdom

The Uneasy Awakening

Artificial intelligence felt like a distant novelty in the early 2020s, a marvel in labs and a fantasy in science fiction. By 2025, AI has permeated every aspect of people's existence, including public discourse, employment markets, educational platforms, search engines, and mental health resources. Fear started to grow, first as a whisper and later as a chorus, while many people embraced the prospects.

Recent surveys show that public sentiment toward AI remains cautious and fearful. About 70% of Americans reported using an AI tool, but nearly half of them believe that AI would result in large job losses, and there is still little confidence in regulatory authorities to protect individuals.[1]

A psychological paradox is marked by this duality: widespread adoption combined with growing anxiety. Despite incorporating AI into their daily lives, people are still very unsure of how it will affect society in the future. This chapter examines the causes of this anxiety, how it affects our thoughts and actions, and how we can respond to it in a way that is resilient and empowering.

The Subtle Integration of AI into Human Life

In contrast to previous technologies, AI does not just enhance human work; it engages with human cognition itself. Each recommendation and forecast mirrors human behaviour and choices, influencing our decision-making.

From a technical standpoint:

- Machine learning models identify complex patterns in massive datasets.

- Predictive algorithms anticipate needs based on past behaviour.

- Generative systems produce content that mimics human creativity and style.

Humans feel a combination of support and introspection on a psychological level. AI has the ability to feel intuitive, almost sympathetic, as though it knows us better than we do. Yet this intimacy poses subtle risks: reliance, cognitive offloading, and the desire to over-rely on algorithmic guidance.

As humans, we must decide whether to use AI as a passive tool or actively foster a cooperative relationship that advances our thoughts while preserving individuality.

A New Relationship with Intelligence

The emergence of AI as a cognitive partner, a working companion that improves decision-making, problem-solving, creativity, and daily productivity rose above all other AI trends by late 2024 and early 2025. Searches exploded for phrases like:

- "How to use AI at work efficiently"
- "AI tools for productivity"
- "AI decision assistant"
- "How to think with AI, not rely on it"

People are looking for amplification rather than just automation. AI is increasingly viewed as an on-demand second mind that can:

- generate options
- model complex decisions
- summarise enormous information sets
- recommend changes
- offer neutral perspectives
- assist with time, thought, and attention management.

However, what the majority of people currently desire, and what this chapter provides, is a useful paradigm for utilising AI in a purposeful, empowered, and human-centred manner. This is the shift we explore: AI is no longer a tool you use; it is a tool you think with.

How AI Enhances Human Thinking

According to psychologists, human cognition is a balance of memory, logic, intuition, emotion, and creativity. AI, on the other hand, offers complementary advantages such as vast memory, quick analysis, structural clarity, and relentless pattern recognition. These two types of intelligence combine to create something greater than either one by itself.

To visualise this, imagine a simple diagram that captures the way these abilities overlap:

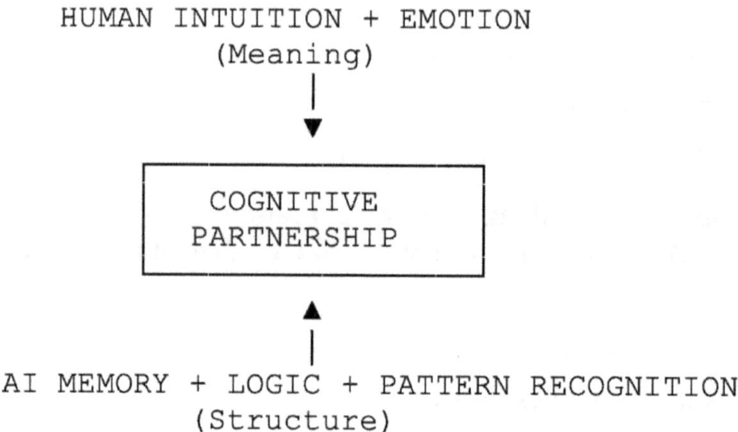

How Cognitive Partnership Actually Works

A natural rhythm emerges when collaborating with AI. It feels like thinking in a more organised and expansive manner rather than following a protocol. This rhythm can be visualised as a loop, reflecting how skilled thinkers handle complexity:

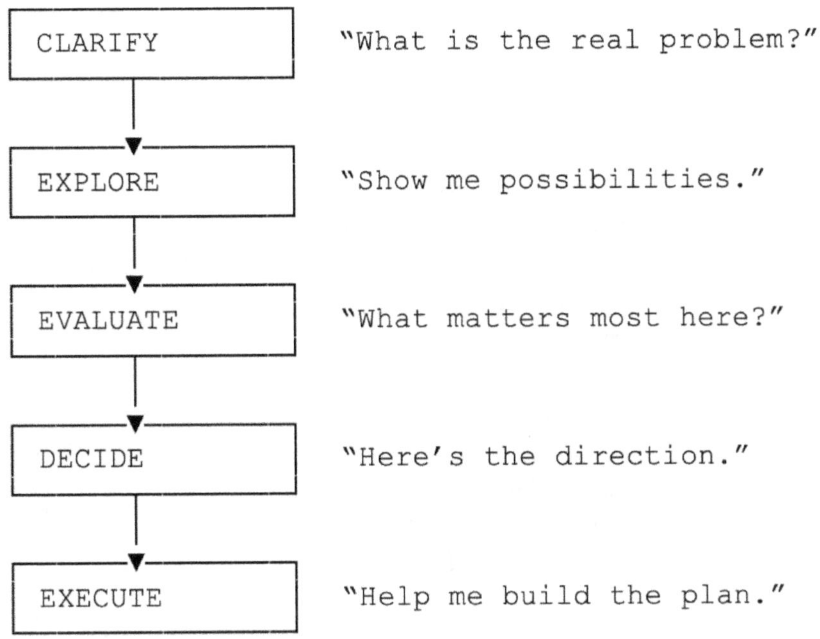

AI helps one movement flow into the next, but human analysis and decision-making will always be at the forefront of it.

Learning Transformed by a Thinking Companion

Learning becomes more humane and visual through AI. For example, when a cybersecurity student asks AI for a learning roadmap, the system cannot only generate a list of topics but also a cognitive ladder:

```
AI-GENERATED LEARNING LADDER
────────────────────────────────────────

Tier 1: Foundations
    • Networking basics
    • Encryption concepts
    • Authentication models

Tier 2: Tools & Practice
    • Firewalls
    • Pen-testing frameworks
    • SIEM tools

Tier 3: Real-World Scenarios
    • Incident response
    • Threat modelling
    • Risk assessment
```

The Future of Cognitive Partnership

The most significant insight of this era is that AI is not becoming a human. Rather, people are getting access to cognitive scaffolding that enhances their innate strengths. By releasing us from mental clutter, the partnership allows us to concentrate on relationships, meaning, ethics, and creativity.

Two kinds of intelligence, one structural and analytical and the other emotional and intuitive, appear to be co-existing in the future. Additionally, when they work together, they generate a degree of understanding and clarity that neither could achieve alone.

To visualise this relationship, here is a diagram of a balanced system rather than a hierarchy or command structure:

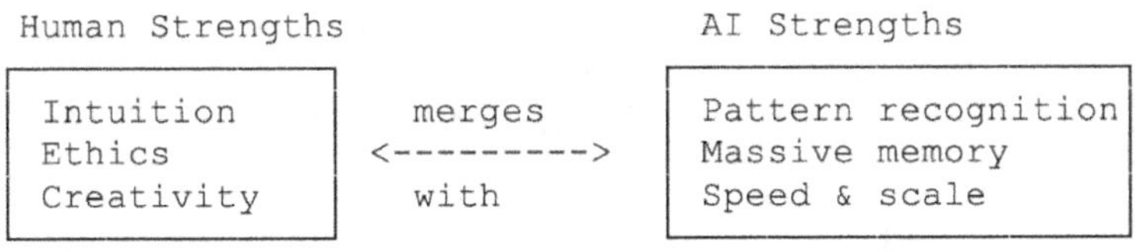

The fundamental idea of the cognitive partnership era is that AI helps humans think more fully rather than thinking for us. And that kind of thinking has the power to reshape everything.

AI Expands Working Memory

Realistically, human working memory is finite as it can only store a few ideas at once. If overburdened, tasks accumulate, priorities blur, and details fall between the cracks. This ability is expanded by AI by keeping a track of:

- dozens of ideas for a project
- timelines, dependencies, and documents
- long-term goals and weekly habits
- meeting notes and themes
- complex plans with many moving parts

In order to free up your brain to do what it does best, which is to connect the dots, feel, envision, and make decisions, AI helps in transforming your memory into a living, changing mind space that stores and organises cognitive load.

AI Accelerates Creative Exploration

Variation, experimenting, and perspective shifts are all essential to creativity. AI is an expert at creating quick iterations, which will enhance human creativity rather than replace it.

AI can be used by a writer to experiment with alternate endings, rethink situations, and explore character arcs. In just a few minutes, a marketer can come up with 20 marketing ideas. Without running into a dead end, a designer can come up with ideas for colour schemes and layouts. AI takes on the role of a spark generator, assisting individuals in transitioning from creative stagnation to flow.

AI Enhances Decision-Making

Every day, humans make hundreds of decisions that could be trivial or life-changing. There is a significant problem with decision fatigue. AI helps by structuring choices, comparing options, and surfacing factors you might not have considered. AI is capable of highlighting strategic trade-offs, identifying decision biases, simulating results, and consolidating vast amounts of data. It doesn't make decisions for you. However, it provides you with a more comprehensive and lucid decision landscape.

AI Supports Metacognition

One of the most advanced cognitive abilities in humans is metacognition, or thinking about how you think. It is the cornerstone of understanding, self-awareness, and personal development. By asking clarifying questions, pointing out inconsistencies in your logic, providing alternative framings, and expressing viewpoints you might overlook, AI enhances metacognition. It helps you observe your own thought processes more clearly, acting like a cognitive mirror.

Staying Relevant in an AI-Driven World

Instead of competing with AI, humans need to evolve alongside it. Staying relevant requires focusing on unique human capabilities such as:

- Contextual judgment: comprehending ethics, culture, and subtleties
- Emotional intelligence: the ability to recognise and react to emotions genuinely
- Innovation and creativity: envisioning what does not yet exist
- Ethical reasoning: solving problems that are beyond the capabilities of technology
- Purpose-driven thinking: coordinating behaviour with purpose, values, and society

Practical Guidance

Here's my advice for what focusing on those human capabilities and staying relevant in the AI world entails:

- Develop abilities that enhance AI rather than compete with it.
- Lifelong learning is crucial; never stop learning and adapting.
- Continue to be conscious of how AI affects social patterns and thought processes.
- Take initiatives in fields where human insight is crucial, such as strategy, ethics, and creativity.

Emotional Quotient (EQ) as a Force Multiplier, Not a Soft Skill

Emotional quotient is often misinterpreted as a "soft" skill, often considered secondary to technical proficiency. EQ actually functions as a force multiplier. A technically proficient individual with low EQ may find it difficult to lead, collaborate, or adapt. By understanding how to analyse outputs, share insights, and apply them effectively, an individual with a high EQ can increase the value of AI.

The consequences of poor EQ increase dramatically as AI systems become more powerful and integrated into large-scale decision-making. This is because miscommunication can now spread at an unfathomable pace, poor leadership can affect entire organisations or platforms, and ethical failures can spread widely before they are identified or fixed. In an AI-driven world, EQ is no longer just a personal skill but a crucial component of risk management.

Where Humans Will Consistently Outperform AI

There are several areas where humans will continue to stand out, utilising emotional intelligence:

- Leadership and influence: motivating people, resolving conflict, aligning teams around shared goals
- Ethical judgment: navigating moral ambiguity and social consequences
- Complex negotiation: understanding unspoken interests, emotions, and power dynamics
- Care-based professions: healthcare, therapy, education, coaching, and social work
- Creative collaboration: co-creating with others, responding to feedback, and evolving ideas emotionally as well as intellectually

AI works best as a collaborator rather than a replacement for human skill in many industries.

AI in healthcare can analyse large amounts of patient data, detect early cancer signals, and detect diseases with immense accuracy, but no patient wants to receive life-changing news from a screen. While doctors utilising AI-based alert systems can detect issues earlier, those integrating medical knowledge with AI insight become augmented clinicians, capable of deeper, more proactive care. AI improves medical capabilities, but people are still essential to healing.

AI is used in law to help with document drafting, case summaries, precedent analysis, and even outcome predictions. However, it is unable to navigate human behaviour, negotiate delicate settlements, comprehend intent, or exercise ethical judgment. Consequently, attorneys are becoming more and more strategists, advisors, and negotiators, which AI cannot replace.

While AI enhances simulations, optimises materials, and automates repetitive operations in the construction, engineering, and manufacturing industries, humans are still responsible for real-world problem-solving, innovative design choices, cross-disciplinary coordination, collaboration, and safety judgments in unpredictable environments. By lowering errors, shortening design cycles, and enabling initiatives that were previously too difficult to undertake, AI speeds up these businesses rather than slowing them down.

The Jobs AI Will Create, Many of Which Don't Exist Yet

The number of new opportunities that AI creates is its most underappreciated effect. As intelligent systems proliferate, entirely new roles are forming around how AI is designed, guided, and applied.

Similar to planners creating a digital city, AI workflow architects design the underlying systems by choosing models, linking tools, and integrating AI into routine activities.

By combining psychology, language, and design, prompt strategists go beyond simple instructions to create multi-step reasoning flows, incorporate ethical and stylistic constraints, and improve AI behaviour.

As positions change rapidly, digital apprenticeship managers combine workforce planning, curriculum design, and coaching to build AI-assisted learning pathways.

One of the fastest-growing professional fields, AI safety, ethics, and governance specialists manage bias, legal risk, misinformation, and unexpected outcomes while supervising large-scale deployments.

Human-AI facilitators will act as a link between humans and intelligent tools by guiding training, settling disputes, and coordinating AI systems with practical processes.

AI-driven climate and sustainability experts use intelligent systems to improve supply chains, energy, and agriculture, generating jobs that promote both environmental resilience and economic growth.

Synthetic media architects shape AI-generated film, audio, virtual environments, and digital identities, driving a creative renaissance similar to the rise of CGI.

Personalisation experts guide AI in tailoring education, healthcare, wellness, and content, blending technology with human coaching and judgment.

Together, these roles reflect a broader truth: AI does not just change work, it creates entirely new ways for humans to contribute.

Increased Entrepreneurship and Reduced Barriers

One of AI's most powerful economic effects is its ability to lower entry barriers. In the past, starting a firm required large teams, specialised knowledge, and significant resources. AI diminishes all three. People with brilliant ideas may now create products, create marketing collateral, investigate rivals, automate procedures, and access global markets with minimal resources.

This democratisation of capabilities is what drives entrepreneurship. More people can launch firms, explore new ideas, and fill previously unreachable niche markets. As these businesses grow, they create a need for human labour in areas such as sales, customer service, design, training, compliance, and leadership. Jobs are generated by opening up new economic activity rather than by maintaining existing tasks.

The Future Belongs to Integrated Intelligence

Humans and machines will not compete in the future. EQ serves as a bridge in this collaboration between human and artificial intelligence.

Individuals who solely depend on technological proficiency could become interchangeable, but those who combine both emotional intelligence and AI fluency become indispensable. They convert knowledge into action, information into choices, and systems into human results. In an era where machines can think faster than we can, humans stand out by feeling, judging, connecting, and caring better than machines ever will.

As intelligence spreads and becomes mechanised, it is ultimately emotional intelligence that keeps the world human.

The fact that AI compels mankind to climb the value ladder may ultimately be its greatest economic contribution. Humans concentrate on purpose while machines manage efficiency. Creativity increases as automation lessens tedium.

In the AI era, economic growth will come from the creation of new employment rather than the preservation of existing ones and from jobs where judgment is more important than repetition. When AI is used carefully, it moves the emphasis from speed for its own sake to deeper understanding, where meaningful contribution is more important than mere conformity. This ultimately shapes a more dynamic, adaptable, and human-centred economy, where growth is fuelled by enabling people to think, create, and make better decisions rather than by completely replacing them.

Endnote

1. Ashley Gold, "Exclusive: Americans Are Using and Worrying about AI More Than Ever, Survey Finds," *Axios*, December 10, 2025, https://www.axios.com/2025/12/10/americans-ai-use-worry-survey.

About the Author

Pankhuri Bansal is an AI and blockchain expert appointed by the United Nations (UN/CEFACT), the International Organisation for Standardisation (ISO), and the founder and CEO of Blockom Consulting. With over 13 years of fintech experience, she advises global organisations and start-ups on emerging technologies and is a frequent speaker at international AI and blockchain conferences. Formerly with JP Morgan's Onyx division, she contributed to enterprise blockchain solutions and has published research papers and standards with both the UN and ISO. She was on the cover page of Entrepreneur Magazine's "35 Under 35" Emerging Tech leaders for 2025 and holds advanced triple Masters degrees in AI, IoT, Economics, Finance & Entrepreneurship from leading UK and US universities.

Email: Pankhuri.m.bansal@gmail.com

Website: www.blockom.net

LinkedIn: https://www.linkedin.com/in/pankhuri-bansal-54097318/

FROM DIGITAL TRANSFORMATION TO INTELLIGENT TRANSFORMATION: LEADING WITH AI AT THE EDGE OF DISRUPTION

By Jason Bowden
Product & Technology Executive, Speaker and Consultant
Atlanta, Georgia

The Pattern We've Seen Before

Digital transformation disrupted how we worked. Remember?

Paper processes became digital workflows. Filing cabinets turned into databases. Phone trees gave way to Teams channels. The disruption felt real. During every system rollout, many of us silently wondered, "If the system does all this, what's my job?" But roles didn't vanish; they evolved. Employees stopped being paper couriers and became data mediators, the glue that makes sense of it all. We went from ripping the sprocket ears off dot-matrix reports to being the "maestros of dashboard KPIs." We weren't replaced; we moved up the value chain.

But here's what it didn't give us: time. Calendars overfilled. Inboxes never quieted. Now we are constrained by redundant tasks that force our navigation across misaligned systems. We are squeezed to the edges of our day with no time for curiosity, judgment, or invention.

We're limited now by capacity, and that is where intelligent transformation becomes the next evolution in the way we will work. This evolution will remove these redundant and repetitive tasks, so we are freed to focus on maturing our irreplaceable human abilities. Don't be mistaken, this chapter is not about techie tool usage, rather it's about how we must design and write narratives that instruct AI how to achieve outcomes in a guarded and ethical manner. So, let's begin the journey to redesign our relationship with work by understanding what's changing.

What Has Changed?

Intelligent transformation is about redirecting our time from operators of repetitive tasks to designers and governors of high-value outcomes. Once again, we are moving up the value chain, but this time the work is about how we frame context, apply judgment, and build the governance that makes AI trustworthy. With that said, the intelligent transformation ecosystem has grown out of a couple of fundamental technological concepts that we must have a basic understanding of to begin crafting our new way of working.

First, generative models (e.g., ChatGPT, Claude, Gemini, etc.) give you natural-language access to essentially all documented human knowledge. Brilliant at synthesis, explanation, and problem-solving. You've probably tested one and been impressed.

The catch is they are PhD grads that are children with no life experience. They lack context, emotional intelligence, and ethical grounding. They haven't lived through hardships, celebrated victories, or felt consequences. Their memory is narrow, and they forget what we said three prompts ago. And they hallucinate by blending fact with plausible fiction, all delivered with total confidence.

As an example, ask a GPT of your choosing to draft a compliance memo, and it might cite a regulation that doesn't exist. Ask it to analyze employee sentiment, and it misses that "thanks for the feedback" is sarcasm.

Why does this matter? Those limits aren't flaws, rather they're our job description. These models must have humans decorate all interactions with detailed context, judgment, ethics, and guardrails. This understanding of what exactly the model can't do tells us why we are irreplaceable.

Second, agentic AI combines the generative model capabilities with the ability to do work given a goal or task. These are the AI workers who can book meetings, reconcile datasets, monitor systems, and escalate issues within a defined level of autonomy.

The catch is they inherit all the LLM limits (hallucination, missing context, no ethical judgment), but now they're acting based on those flaws. An agent doesn't just suggest a wrong answer. It executes on it. It might update a record incorrectly, send an email with fabricated details, or approve a transaction that violates policy, all the while appearing deceptively confident.

Let's analyze a real-world example using the "good agent" and the "ugly agent" paradigm to illustrate these potential impacts: Imagine Dana, a VP of operations at a national property firm, wants to speed up tenant approvals and launches an agent to define a "rental risk score" across hundreds of applicants. The agent's goal: "Rank applicants by a rental risk score, so staff can approve the safest choices." What makes this dangerous? That agent could easily derive a rental risk score using unethical reasoning.

The Ugly Agent: The agent's rental risk score shows up in the leasing system like any other field. At first, it feels like a win. Approvals are faster. Backlog drops. Then, six months in, a couple of local fair housing agencies receive a rash of complaints about the company's denials. They start an investigation and discover that many applicants with solid income, rental history and references are being denied for no good reason. Further investigation reveals that many of the applicants unfairly denied happen to have lived primarily in lower-income neighborhoods. Eventually the investigation gets headlined by news agencies: "Screening Algorithm Unfairly Denies Renters Across Hundreds of Applications." Lawsuits follow. Regulators step in. The firm's brand takes a hit that won't wash off with a press release.

The Good Agent: Same desire for speed. Same idea of scoring. But Dana's team treats the agent like a powerful administrator, not the decision maker: A human reviewer is required and documented before any rejection decision is made. Why does this matter? Agents can amplify both capability and risk! That PhD five-year-old isn't just answering questions; now it's making policy decisions. Without clear boundaries, the right ethical context, and human review, it will confidently do the wrong thing on a disastrous scale.

Clearly the lack of these human-centric qualities is the fundamental weakness of AI for now and a long time to come. With that said, significant human advancement in virtually every field will be unlocked by those who can design valuable agents that are trustworthy and safe. So, let's begin to shift our mindset into a new frame where our role moves up the value chain. This is where a lot of people freeze, and they feel the shift coming but don't know where to grab it. The way is straightforward! Master how to get quality output first by treating prompts as a designed method, not a casual question.

Getting Quality Output

During digital transformation, the advantage went to the people who could turn "What's going on here?" into a clean report or dashboard. Now, our next step up the value chain is to learn how to write prompts that give generative AI sharp context, scope, and boundaries. When we do that well, we get output that is relevant, more trustworthy, and less likely to hallucinate. The structure of the prompt should follow this basic model:

- Goal: What does "done" look like?

- Persona: Choose the perspective for the right tone, depth, and assumptions.

- Inputs and Context: Be strategic and aware about exactly what information it has and what each piece is.

- Process: Give it a simple checklist of steps.

- Constraints and Boundaries: Set limits: scope, style, thresholds, and "never do X."

- Output Format: Be strategic and exacting about how it should shape the reply format, so the reply can be used without editing.

- Review and Uncertainty Rules: Decide how it should behave when data is missing, ambiguous, or risky.

Once we can do this reliably, the next challenge is to clearly identify where agents make sense to automate tasks and even entire workflows.

Finding Agentic Opportunities

During digital transformation, we looked for processes that were repeatable enough to automate and standardize. Places where we could eliminate a paper-driven process or turn a mess of data into a usable dashboard. Intelligent transformation is the same hunt, just at the next level. Once we can get quality output from prompts, the next move is to spot where an agent could safely take over parts of the redundant work. Learning how to define tight scope is the best way to start. Think of small, simple, repetitive tasks with clear upside value:

1. The goal is clear. We can concisely finish the sentence: "The agent's job is to _______."

2. The value proposition is clear and measurable in dollars or time saved.

3. Everything happens digitally. The task uses emails, files, systems, or websites. No physical work is needed.

4. It's repetitive. People do it often and in roughly the same way each time and automating it would free valuable time.

5. Mistakes are fixable. If something goes wrong, it can be undone, edited, or rolled back.

6. We can write simple rules for it. There is not a huge variance in the possible outcomes, and we can imagine the risk and what could go wrong.

7. It needs reasoning but not authority. Analyze data, compare numbers, summarize content and perform routine actions based on findings. Stay clear of anything that has a risk of impact on reputation, trust, ethics, or emotions.

Sticking to these concepts helps us spot good agentic candidates instead of forcing AI into the wrong places.

The next question is how that agent should work. When it wakes up, what it sees, how it thinks, and what it's allowed to touch.

Be an Agentic Designer

During digital transformation, our work was to make systems fit the way the business ran by mapping processes, writing SOPs, and creating audit controls. Designing an agent uses the same muscle, only now it is pointed at a new kind of teammate. We would never hire a person, give them full system access on day one, and expect good outcomes. We define their job, limit access at first, and add checkpoints until trust is earned. Agents need the same structure. We capture that structure in a narrative written in detailed, plain language that describes the agent's role and function. The simplest way to start is to imagine a "day in the life of the agent" and spell out, step by step, what wakes it up, what it sees, how it thinks, what it produces, and where humans step in. The outline below is a practical template for that agentic narrative, with concrete examples.

Triggers: Think of the trigger as the moment you tap on the agent's shoulder and say, "Okay, now it's time for you to work." If we get the trigger wrong, everything else is at risk. If we get it right, the agent feels natural and useful. Here are some common trigger types with examples:

- Human-Initiated Triggers: Project Manager clicks "Summarize this meeting."

- Event-Based triggers: A CRM Lead moves to "Qualified."

- Time-Based Triggers: "Every morning at 8 a.m., summarize yesterday's incidents."

- Conditional Triggers: "If response time on tickets exceeds our SLA, alert and summarize."

Inputs: This is everything we hand an agent and say, "Do your job based on this information." Here are key types of inputs and examples:

- Primary Task or Object: support ticket, contract document, meeting transcript

- Immediate Context: message in the support ticket, prior contract version, last meeting schedule

- History and State: trend of tickets, past purchases, prior human overrides

- Rules, Policies, and Thresholds: no PII data, flag missing data for human review

Thinking and Reasoning: This part is our agent's operational handbook. How should the agent work through the problem, and what actions is it allowed to take in what order? Here is how to structure the thinking narrative:

- Checklist: "For each new support ticket: 1) read the message, 2) detect topic, 3) suggest a reply, 4) tag urgency."

- Triage and Branching: "If the ticket is about billing, route to billing workflow; if it's about login, route to access workflow."

- Propose, Critique, Refine: "Draft a variance explanation → re-read the numbers to catch contradictions → shorten and simplify for executives."

- Tool Usage: "Look up customer history in the CRM → find the billing status."

- Human in the Loop (HITL): "Draft email replies and update suggestions but require a manager to approve anything prior to sending."

Outputs: Good thinking only matters if it leads to the right follow-through. Once the agent has made sense of the situation, the real risk and value sit in what it does next. Designing actionable outputs is critical for describing how an agent does its task but collects the required human review and approval.

- Draft Outputs: "Generate a reply email and a proposed subject line, but don't send it, just show them to the user for editing."

- Auto-Apply Output (use in low-risk scenarios only): "Automatically tag tickets with a topic and priority label then move them into the triaged queue."

- Mixed Mode (Auto + Escalate): "For refunds under $50, draft and apply the refund; for anything higher, prepare a recommendation and send it to a supervisor for approval."

Guardrails: Even with good thinking and sensible outputs, an agent still needs specific boundaries called out that clearly set limits on what it can see, touch, and decide. Guardrails are those boundaries, and they keep the agent predictable and safe. It is critical to have a specific guardrail section in your narrative that contains these concepts.

- Data Boundaries: "You can read anonymized transaction data and ticket text but never see full credit card numbers or private notes."

- Human Reviews: "Any outbound message to a customer, or any change over $500, must be reviewed and approved by a person."

- Fail-Safes and Safe Defaults: "If key data fields are missing or discrepancies exist, don't guess, don't drift! Flag the item as REQUIRES REVIEW and STOP."

- Scope and Permissions: "You may create draft tasks and add tags, but you may not delete records or close tickets."

With all these narrative elements in place, we have a fully designed agent that can be tested, validated, and put into production.

Now it's time to govern that agent's operations.

Be an AI Governor

Designing an agent is only half the story. Now the mindset becomes who can use it, who reviews and approves, how do we monitor for drift, and how do we roll it back if something goes wrong? We become AI governors that focus on observability, ethical drift, guardrail adherence. Mastering the following concepts will keep agents' operations on-track, so value is reliably achieved with trust and safety.

- Access and Role Boundaries: Determine who can invoke the agent, who must review, what data it can access, and how it can be updated.

- Agent Risk Classification: Define potential blast radius and mitigations.

- Review Ownership and Approval Chains (Human Oversight): Define where humans must stay in the loop, especially for external, financial, or reputational impact.

- Audit Trail and Traceability: Keep a record of prompts, inputs, outputs, and human edits, so you can see how a decision was produced.

- Versioned Methods (Prompts as Assets): Treat prompts, workflows, and data connections like code (versioned, labeled, and change-tracked).

- Monitoring and Guardrails: Watch how the agent behaves over time and set thresholds where it must slow down, stop, or alert someone.

- Feedback and Improvement Loops: Make it easy for users to say what worked and what didn't and feed that back into the next version.

- Alignment: Ensure and measure how policies are aligned to KPIs (override rates, error rates, financial impacts).

In practice, agent governance shouldn't be invented from scratch. We can lean on several existing governance organizations for guidance: NIST for risk framing and control functions, ISO for management-system discipline (policies, roles, continuous improvement), and COBIT for the executive view of ownership, alignment, and accountability. Together, they give you a language to treat agents as governed systems: scoped, controlled, monitored, and tied back to real business objectives.

Leading with AI at the Edge of Disruption

Digital transformation rewarded the people who could bridge the gap between "what the business needs" and "what the system can do." We already know that game: turn a loose question into a query, optimize the system workflow, make a pile of data into a dashboard. Intelligent transformation is the sequel. Now the edge goes to people who can design the tight narrative, not just drive the tools.

Our skill set shifts from:

- turning messy questions into flashy dashboards and slide decks to engineering prompts that give quality outputs

- mapping processes and writing SOPs to designing valuable and trustworthy agentic workflows

- managing access, tickets, and change logs to defining a governance narrative for agentic workloads

This isn't about any of us becoming expert AI coders. It's about taking the skills we already built in digital transformation and refining them for the next wave: discovering agentic value, designing agent narratives, and operating them with trustworthy governance. That is how we, collectively, lead with AI at the edge of disruption.

About the Author

Jason Bowden is a seasoned leader in data, analytics, and AI transformation with more than two decades of experience, leading organizations to turn data into meaningful, measurable business outcomes.

He is recognized for his ability to break complexity into incremental, value-driven delivery, ensuring organizations see tangible progress while building toward long-term strategic capabilities. He is a champion of stakeholder alignment, helping executives, engineers, operators, and IT teams converge around shared goals, trusted outcomes, and practical roadmaps.

Jason is a passionate advocate for adoption; he believes technology only matters when people adopt it. He focuses on education, advocacy, and change management to turn adoption into real momentum, helping organizations think, operate, and innovate more effectively with data and AI.

Email: Jason.Bowden@nexsage.ai

Website: www.nexsage.ai

LinkedIn: https://www.linkedin.com/in/jasonbowdenprofile/

CHAPTER 5

FROM MACHINE VISION TO AUTONOMOUS SYSTEMS

By Shang-Wei Chao
AI Consultant and Educator
Taipei, Taiwan

It is not enough for machines to be intelligent; we must ensure they are aligned with human values.

—Stuart Russell

Breakthroughs in Machine Vision Technologies

In an era of rapid AI evolution, machine vision powers autonomous systems and humanoid robots, transforming industries through data-driven intelligence, ethical collaboration, and systemic reinvention. This shift demands visionary leadership to harness opportunities while managing risks for sustainable competitive advantage.

Machine vision is rapidly evolving from simple image recognition into a core engine driving business intelligence and autonomous systems. For managers and technologists, understanding its cutting-edge advancements is no longer an option but a cornerstone of future strategy. Trends in recent years highlight two key technologies—Vision Transformers (ViT) and Generative Adversarial Networks (GANs)—that are enhancing machine "sight" in unprecedented ways. Combined with privacy-preserving Federated Learning, these technologies are unlocking a new chapter of digital transformation in business applications.

Traditional Convolutional Neural Networks (CNNs) examines an image piece by piece, extracting hierarchical features from local patterns to assemble the complete picture. In contrast, Vision Transformers (ViT) takes in the entire image to understand the relationships and global context between all its parts. This ability to see the "big picture" allows ViT to achieve far greater accuracy than traditional models in tasks requiring high-level contextual understanding.

More efficient ViT models can now be deployed directly on edge devices, enabling real-time, zero-latency decision-making. This on-device intelligence allows systems to instantly identify subtle anomalies or recognize critical patterns from visual data, significantly boosting operational efficiency and service quality across a wide range of applications.

Data is the fuel for AI, but acquiring high-quality, diverse data in the real world, especially for rare events like unusual anomalies or edge cases across various domains, is both expensive and time-consuming. Generative adversarial networks (GANs) offer a revolutionary solution. Comprising a "generator" and a "discriminator" that compete and co-evolve, GANs can produce synthetic data that is virtually indistinguishable from real data.

The business value is immense. A manufacturer no longer needs to wait months to collect enough defect samples. Instead, they can use GANs to generate thousands of images with subtle defect variations, rare medical anomalies, or unusual environmental conditions in a matter of hours. This synthetic data can then be used to train a highly perceptive AI model (e.g., ViT) for quality control, diagnostics, or scene recognition. This not only dramatically reduces data acquisition costs but, more importantly, enables the AI model to handle a wider range of unexpected scenarios, enhancing system robustness.

Another issue is when data involves user privacy (e.g., customer behavior in retail) or high confidentiality (e.g., medical imaging), traditional centralized AI training methods face significant risks and compliance challenges. Federated Learning emerges as an ingenious solution to this dilemma. Under this framework, instead of centralizing data, the AI model is "dispatched" to local devices or servers where the data resides to be trained locally. Ultimately, only the updated model parameters, containing no raw data, are sent back to a central server for aggregation.

Federated Learning aligns with the "data collaboration" strategy promoted in business literature, enabling organizations, even competitors, to jointly train superior AI models without revealing sensitive data. For instance, multiple hospitals could co-develop a state-of-the-art cancer detection model without sharing any patient's private information. This "benefit without sharing" paradigm provides a secure and innovative path for data-driven digital transformation, enabling companies to build a unique, collaborative competitive advantage in the age of AI.

Integration and Applications in Autonomous Systems

If ViT and GANs are the technologies that give machines their "eyes," then autonomous systems are the "brain" and "body" that these eyes serve. The true power of machine vision is realized in how it enables physical agents like vehicles, drones, and robots to navigate, decide, and interact autonomously.

Modern autonomous systems are undergoing a paradigm shift from "map dependency" to "world understanding." Traditional autonomous driving has heavily relied on high-definition (HD) maps and LiDAR, akin to running on rails. However, vision-based navigation allows a system to operate like a human, primarily "seeing" and interpreting the environment in real time through cameras to make judgments. This grants the system the ability to operate in unknown or dynamically changing environments.

Technically, this is powered by advances in "developmental machine autonomy." AI is no longer just executing pre-programmed rules but is building an internal model of how the world works through continuous learning, achieving "self-directed interpretation." For technologists, the greatest challenge lies in processing the massive stream of visual data (thousands of frames per second) and closing the perception-planning-control loop in milliseconds. This places extreme demands on both algorithms and computational hardware.

The Robotaxi Industry Snapshot

The robotaxi industry serves as the ultimate proving ground for the maturity of autonomous systems. In recent years, key players showcase distinct technological paths and business strategies. Tesla leverages real-world data from its massive fleet for the most aggressive "vision-only" strategy, relying solely on cameras without sensors like LiDAR. This path aims to eliminate safety drivers for unsupervised rides, with a core model transforming existing vehicles into revenue-generating robotaxis via over-the-air software updates.

In contrast, Waymo (owned by Google), the industry pioneer, has launched commercial driverless services in several US cities. It uses a conservative "multi-sensor fusion" approach, integrating cameras, LiDAR, and radar for redundancy and safety at higher costs, building trust with regulators and the public. Similarly, Cruise (backed by GM) and Zoox (owned by Amazon) target dense urban environments, with Cruise drawing from extensive testing in areas like San Francisco and Zoox designing bidirectional vehicles without steering wheels to redefine urban mobility.

Meanwhile, Uber acts as a platform aggregator, partnering with tech providers like Waymo to integrate robotaxi services into its ride-hailing ecosystem. This light-asset model enables rapid scaling without heavy hardware investments, mitigating risks of developing autonomous technology from scratch.

For managers, adopting autonomous systems is not just a technological upgrade; it's an opportunity to redefine operational models. However, a successful strategy is not about incremental improvements but about systemic change.

First, leaders must confront the risks of AI bias and failure. An AI trained in sunny California may perform poorly in a Moscow snowstorm; similarly, an AI optimized for right-hand traffic in the US may perform poorly in left-hand traffic environments like Japan; a single high-profile accident can destroy public trust, leading to severe brand and legal crises. Therefore, establishing a robust "trust and safety" team for continuous, cross-scenario testing and validation, while maintaining transparent communication, is a critical priority for any adoption strategy.

Second, the core objective is to maximize human-AI collaboration efficiency. The value of autonomous systems lies not in completely replacing humans, but in freeing them from repetitive and dangerous tasks. This allows people to transition into roles as supervisors, strategists, and exception handlers. Organizations need to redesign workflows and invest in employee training, teaching teams how to work effectively with their new AI "teammates" to achieve a synergy where 1+1 > 2.

Advances in Robotics: Physical AI and Humanoid Robots

We are witnessing a fundamental paradigm shift: Robotics is evolving from traditional, rule-driven automation to data-driven "physical AI." The humanoid robot, as the ultimate general-purpose form factor, is moving from science fiction into real-world factories and warehouses, and even everyday life.

"Physical AI" completely transforms the traditional, pre-programmed paradigm. These robots no longer rely solely on fixed, human-written instructions. Instead, they learn generalizable skills from vast datasets, spanning both simulation and the real world. This "data-driven" approach allows robots to understand physical common sense (e.g., how objects interact, what constitutes a stable structure), enabling them to dynamically adapt to changes and handle exceptions. Generative AI plays a key role here, allowing a robot to autonomously "generate" solutions and action sequences for novel tasks it has never seen before, achieving true flexibility.

The convergence of two key technologies marks an inflection point for humanoid robots. Hardware breakthroughs, such as Boston Dynamics' all-electric Atlas, shift from bulky hydraulic systems to quieter, cleaner, and more precise electric actuators. This enhances dynamic performance and force control while reducing operational costs.

A flexible body alone is insufficient. Advanced AI integration, such as Figure AI's Helix model, Google DeepMind's robotics models, and Tesla's Optimus, equips robots with intelligent "brains." These enable understanding commands via vision and language, complex reasoning, and translation into fluid, precise physical actions.

For managers, the business logic for deploying humanoid robots is becoming clear. This is not just automation; it is a strategic investment.

The first layer of logic is cost-parity and workforce augmentation. As the manufacturing cost of robots rapidly declines and human labor costs rise, the total cost of ownership for humanoids in many roles is fast approaching or even falling below that of human labor. They can take over the "dull, dirty, and dangerous" jobs, solve labor shortages, and free up human employees for more creative and strategic high-value activities.

The second, more strategic layer is flexible automation and operational resilience. Traditional automation is rigid; a production line designed for Product A cannot easily be switched to make Product B. A humanoid robot, however, is software-defined hardware. When market demand or the supply chain shifts abruptly, a company can rapidly redeploy these robots to new tasks via software updates. This unprecedented flexibility provides a powerful weapon against uncertainty, enabling businesses to build more resilient production and logistics systems. Tesla's plan to deploy its Optimus robots at scale in its factories is driven by this potential, aiming to create a "flexible factory" that can quickly adapt itself.

AI Adoption and Transformation in Industries

A more critical question remains: How can these powerful technologies be translated into a sustainable competitive advantage? The answer lies not in piecemeal technology procurement, but in a top-down business revolution. A successful AI transformation is a shift in mindset, demanding that companies move from "incremental improvements" to "systemic reinvention," steered by leaders with an "AI-first" mentality.

Many companies, when adopting AI, treat it merely as an IT upgrade to optimize a single process, such as using an algorithm to improve click-through rates in marketing. While these "point solutions" can yield short-term benefits, they fail to build a long-term competitive moat. AI's greatest value is unleashed through the network effects it creates as a "system."

Consider autonomous driving. The true market disruption comes not from a single driver-assist feature, but from the entire "transportation-as-a-service" system. Companies like Tesla and Waymo are not just developing cars; they are building a complex ecosystem of data, algorithms, hardware, operations, and business model. Once this system matures, its efficiency and cost advantages will completely upend the traditional auto sales and taxi industries. This systemic approach is the true path to victory in the age of AI.

A systems approach creates entirely new business models in autonomous driving. Beyond robotaxis, two players illustrate how AI targets specific industry pain points. Aurora (Freight Logistics) skips the complex passenger market to focus on long-haul logistics. Its "driver-as-a-service" model charges per mile, addressing driver shortages, high labor costs, and downtime from required rest. Rather than selling trucks, Aurora provides efficient, economical freight capacity. May Mobility (Public Microtransit) addresses urban transit's "first-mile, last-mile" issue. Its "autonomy-as-a-service" platform partners with ride-hailing like Uber and public systems for short-trip shuttles. This B2B2C, light-asset model integrates quickly into existing networks as a complement, not a disruptor.

Technology alone does not create value; it blossoms only with the right leadership and culture. A visionary AI leader must possess three core traits:

1. View Data as the Lifeblood of Intelligence: They recognize that in the machine learning era, data is not just an asset. It's the raw material for emergent intelligence. This means architecting a "data flywheel" where high-quality, diverse datasets continuously feed AI models, iteratively improving them. They break down silos not merely for efficiency but to enable cross-domain insights, such as using customer interaction data to refine supply chain predictions, while embedding ethical governance to address biases from the ground up.

2. Embrace Comprehensive Ecosystem Thinking: Beyond optimizing isolated processes, they reimagine the business as an interconnected AI ecosystem, where technologies like edge computing and Federated Learning converge to create resilient, adaptive systems. This involves building interdisciplinary teams of engineers, business experts, legal experts, and domain experts to simulate and stress-test AI integrations across the value chain, turning potential disruptions (e.g., supply shocks) into opportunities for proactive innovation.

3. Cultivate Collaborative Human-AI Partnerships: They grasp that true AI power lies in augmentation, where humans provide creativity and ethical oversight while AI handles scale and precision. Drawing from cognitive science, they invest in "hybrid intelligence" programs, reskilling employees in prompt engineering and model interpretation to form feedback loops that evolve both people and algorithms. The result: Organizations where AI

acts as a cognitive enhancer, amplifying human potential and driving exponential productivity gains.

Take retail as an example, an AI-first leader isn't satisfied with just a recommendation engine. They push for an integrated system: using machine vision for automated checkout and smart inventory management, leveraging federated learning for ultimate personalization with privacy, and using AI prediction models to optimize the entire supply chain. This is the leap from technology application to business transformation.

Challenges and Risk Management

As businesses embrace the immense potential of AI, they must also confront its significant attendant risks. A single high-profile failure of an AI system can be far more damaging than a traditional IT system malfunction. It not only causes financial loss but can also destroy decades of brand trust overnight. Therefore, a robust framework for challenges, ethics, and risk management is not a "brake" on the journey of AI transformation, but rather the essential "guardrail" that ensures a company can proceed steadily and far.

AI risks primarily arise from at least two sources:

System Failure: Unlike traditional software bugs, AI failure modes are more complex and unpredictable: An autonomous vehicle might misidentify an obstacle due to a rare light-and-shadow pattern or a medical AI's diagnostic accuracy could decline over time due to data drift. Because these systems are granted significant autonomy, the consequences of their failure can be catastrophic, triggering public panic and severe regulatory action.

Embedded Bias: If training data contains historical societal biases (e.g., regarding race or gender), the AI will faithfully learn and amplify them with alarming efficiency. A recruiting AI trained on historical data might systematically filter out resumes from female engineers; a facial recognition system undertrained on certain ethnic groups could lead to false identifications. This is not only a serious ethical issue but also a direct legal and business risk that can alienate large market segments and damage corporate credibility.

In the face of risks, experts recommend a strategy of "gradual revolution," avoiding high-risk, "big-bang" deployments. The core idea is to build trust and system robustness progressively by strengthening human-AI collaboration.

The Ultimate Principle: The Rise of Hybrid Intelligence Systems

Regardless of how technology evolves, the most efficient and robust model of the future will be the "Hybrid Intelligence System." This is a symbiotic model that combines machine autonomy with human oversight. The goal of this model is not to achieve 100% "unmanned" automation, but to achieve the "maximization of human value."

This hybrid model perfectly resolves the conflict between technical scalability and ethical business deployment. Machines provide the scale, while humans ensure the direction is correct and remain accountable for the outcomes. The companies that can master how to design, manage, and optimize these hybrid teams the fastest will gain the most durable competitive advantage in the age of AI. AI and robotics will reshape every industry profoundly, and victory will belong to the visionaries who start planning for a future of human-machine symbiosis today.

About the Author

Paul Shang-Wei Chao has been a research scientist and chief consultant with BroadMission Corporation since 2016. In 2019, he joined the United Nations Centre for Trade Facilitation and Electronic Business (UN/CEFACT) as a domain expert. He also serves as a TT&L WG domain expert for the Asia Pacific Council for Trade Facilitation and Electronic Business (AFACT). In 2021, he served as an industry domain expert with the Bureau of Industry, Ministry of Economic Affairs, Taiwan. He is currently a PhD candidate in the field of electrical engineering at National Taiwan University, Taipei, Taiwan. His current research interests include intelligent computing, deep learning time series, the smart internet of things (IoT), sensor systems, medical intelligent computing, computer architecture and embedded systems, and machine vision.

Email: sw.chao@broadmission.com.tw

Website: www.paulchao.org

LinkedIn: https://www.linkedin.com/in/paul-chao/

DESIGNING WITH INTELLIGENCE: AI-ENHANCED ENGINEERING FOR THE AUTOMOTIVE INDUSTRY

By Christiano Gerhart Dittrich
Mechanical Design Engineer
Joinville, Brazil

Not every path is meant for every traveler.

—Unknown

Due to the exponential growth of practical applications and the awareness that we stand at the beginning of a radical transformation of humanity, it is useful to step back and stir in all of us the will to invest some extra effort to grasp a wider panorama of the reality now unfolding. We must discuss, responsibly and without fear of contradictions or perplexities, the likely or hypothetical dangers that artificial intelligence presents.

Honest judgment about AI depends on the variety of viewpoints, including contradictory ones, that we should adopt as "ours," provisionally, until the moment we retain what we deem worthy of being welcomed. (This is what Edmund Husserl, the German philosopher who founded phenomenology, called the suspension of judgment (*epoché*).) Without this openness, what could be a refreshing discovery becomes something "incredible," in the literal sense of not believable, discarded and set aside, at least as a hypothesis.

Simulated Thinking

Although most of us know, even if intuitively, that intelligence is not the mere ability to solve math problems, present statistical probabilities, or store every word in the dictionary, the term "intelligence" in "artificial intelligence" is used by convention in the scientific community since the Dartmouth Proposal (1956, McCarthy).

Take the following two statements:

- "[...] imitate a human being by answering questions sufficiently well to deceive a human questioner for a reasonable period of time."

- "According to Turing, a computer could be described as intelligent if it could deceive a human into believing that it was human."

The first is a summary of the so-called "Turing criterion" found in Shannon and McCarthy's preface, which synthesizes Turing's proposal. The second is an indirect attribution to Turing by Person and Graesser; it is not a literal quotation of Turing, but a poorly made interpretive paraphrase—ultimately a distortion, whether intentional or not, far from the original referent. A literal quotation from Turing is: "I propose to consider the question, 'Can machines think?'"

One should not be called intelligent for the capacity to deceive another person, but for the capacity to learn the truth, either from others or from reality itself in its endless nuances. Since a computer is not someone but something, it becomes even more difficult to ascribe intelligence to it. An object—something—does not deserve merit; a person does, by virtue of quality, effort, or conduct.

"Intelligence lies in the achievement of the end, not in the nature of the means employed. And if the end of the means of knowledge is to know, and knowledge is fully knowledge only if it knows the truth, then the definition of intelligence is: the power to know the truth by whatever means." In essence, intelligence is the disposition to choose the best end: truth beyond appearances. Having a powerful model at hand does not mean having intelligence, but having intelligence may well mean choosing to use the model to refine one's own intelligence in a feedback process.

If the model improves qualitatively only through human intervention—without the slightest interference from a ghostly entity hidden in some metaphysical manner manipulating the simulacra of neural networks—then what we face is a simulation, exceedingly fast—bizarrely fast—of human thoughts.

Therefore, based on what has been laid out so far, if we are dealing with a simulation of cognitive processes, the most adequate term to describe this technology would be: simulated thinking. I will continue to use the currently conventional term artificial intelligence as a figure of speech.

Between Sages and Esoterics: Zoom Out First

In this first retroactive movement, in search of a more holistic view of our present context, as powerful as it is dangerous, we must consult the sages, beginning with Plato. Recognized as an advocate of science as necessary and universal knowledge, he divided the cognitive faculties: the highest level is *episteme* (composed of *dianoia* and *noesis*). *Dianoia* concerns demonstrative knowledge through pure thought: mathematical entities, invisible and insensible, with more resistant and more permanent consistency than living beings. *Noesis* (from *nous*) is the dialectical apprehension of first principles (the *Forms/Ideas*—eternal and immutable archetypes). Finally, *nous*, inspired by contemplation of mathematical forms, ascends beyond them, what we today translate as spirit, intellect, and intuition; that is, the eternal *Forms/Ideas*, the first principles.

For Aristotle, one must employ dialectic and the confrontation of hypotheses to verify the validity of first principles, articulated in sequence within coherent logical discourse; that is, science does not deal with separate forms but with natural substances and their causes; it starts from principles (grasped by *nous*) and proceeds by demonstration.

Francis Bacon, a forerunner of the Scientific Revolution at the dawn of modernity, was an archetypal figure in the shift in what qualifies as science, influencing our shared worldview. He crafted a method saturated with rules, beginning from observables disarticulated from first principles and, in my view, inaugurated the overvaluation of mathematics. Thus he introduced the risk to describe a phenomenon with mathematical perfection at a level of abstraction but far from the true cause for want of broader contextual meaning, despite experimental validation. "If you torture the data long enough, they will confess to anything."

His work is not fully consistent with his advice, for it is chiefly programmatic. Moreover, his alleged involvement with esoteric secret societies presupposes participation in, or valuation of, symbolic rites. His scientific-institutional utopia, New Atlantis, is often read as the literary expression of that horizon.

My critique, however, is not directed at his esotericism—even if the covert operation of esoteric societies is intriguing and, in the case of secret ones, at least questionable—nor, for that matter, at the study of symbolism, but at incoherence. From this angle, Bacon's aim seems to have been to lead us to think as he wanted, not necessarily as he truly thought. Entering the rabbit hole is imprudent.

There are other biographies with similar tensions. My aim, however, is not to produce a treatise on the history of science, but to illustrate a fact: Despite the sages' legacy, the extraordinary technological advance, and admirable scientific contributions, many current problems may also stem from our acceptance of vain ideologies or philosophies which—far from making a better world—

bear a record of social degeneration, seek power, and leave a trail of mortality. In this line, attributing absolute historical purity to science or to the scientific class is naïveté, naïveté that sometimes serves artful financiers without our noticing the consequences.

The Articulator of Principles

At times we think through a project and reach no truly good solution; at other times, in moments devoid of conscious thoughts about it, a sudden intuitive flare brings the solution.

At the end of the project, based on your experience, theoretical knowledge, and practical participation, you know which aspects of the product were the best delivery possible given the resources, variables, and expectations, as well as which aspects could be better. You are (or should seek to be) certain of this. Your degree of certainty may vary; what truly changes is how we face reality as it presents itself and the extent of our exposure to true facts.

This is crucial for engineers seeking the best solution. The specialist-generalist—who articulates and integrates techniques of different natures into a single solution and possesses "digital empathy" uses AI efficiently and synergistically, thanks to a broader repertoire, and, thus, is better positioned.

If, however, one clings to the particular case, prioritizing the specific over the whole, failure will be miserable.

At the center of the process stands the self-aware subject. It is the engineer, responsible for the proper use of the AI model and for interpreting results, not AI itself.

This approach is particularly critical in regulated industrial domains—such as pressure vessels and process equipment—where design assumptions become effectively irreversible once fabrication begins or demand substantial resources to correct.

AI technology "at my fingertips" is beautiful; enthusiasm is inevitable. My first use was with GPT-4.5 during the pre-design phase of a large rotary jig/fixture, when I needed to estimate the maximum deflection of the main body before CAD modeling and before FEA verification. My goal was to reduce computational cost from new prototypes and from excessive simulations.

To compute the second moment of area of a section composed of multiple structural shapes, I built an AI assistant grounded in the parallel-axis theorem. Then I integrated another assistant focused on beam analysis by Timoshenko theory.

When the numerical validation in the FEA module produced results within tolerance relative to classical engineering integrated with AI, I moved forward, confident in the consistency of the outputs and the adequacy of the elements and inputs used in the FEA simulation.

Today's AI-oriented hardware architecture does not single-handedly sustain large-scale numerical problem-solving typical of FEA on CAD prototypes with many elements and/or complex simulations (e.g., non-linear); however, it is quite useful in post-processing. Despite this limitation, with appropriate tools, AI becomes a powerful ally to make dense, existing classical-engineering material operational.

In this sense, my engineering business, Dittrich Eng, intends to act. My goal is to gain the expertise to implement a custom informational and normative nucleus for engineering centers: a multi-code platform that combines RAG (retrieval-augmented generation) with verifiable citations, programmed to issue standardized calculation memoranda, with auditable audit logs and traceability.

This effort also extends to applied-AI initiatives for structured communication, including the TableHub.io platform, designed to generate professional texts that guide delicate situations with clarity, while preventing conflicts and reducing legal exposure caused by poorly calibrated or poorly documented communication.

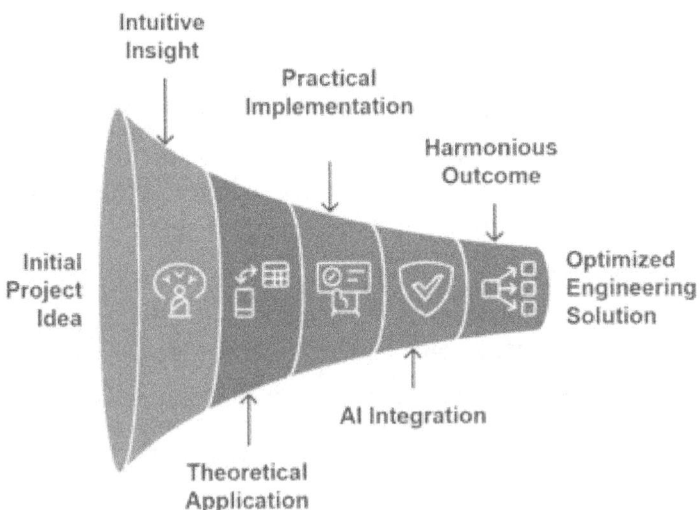

Linguistic Repertoire and Experience: A Second Zoom Out

With the advent of LLMs, it becomes evident that linguistic competence and experience—that is, repertoire—will better position companies and professionals in the market.

If the limits of language define people's place in the world, what of the vast portion of humanity, alas, including "functionally illiterate" university students, who lack communicative competence? I wonder how I survived a school shaped by social engineering (per Pascal Bernardin, *Machiavel pédagogue ? — ou Le Ministère de la réforme psychologique* [1995; "Machiavelli, Pedagogue?—or the Ministry of Psychological Reform"]): the shift from intellectual content to behavioral/attitudinal competencies using goal-based evaluations and psychosocial influence techniques (conformism, group norms, cognitive dissonance, etc.).

While everyone has a duty to self-educate, this silent cultural revolution—anti-democratic and totalitarian—manipulates the masses into submission to the ruling class, preventing many from escaping idiocy on their own.

I do not judge an expanded repertoire necessary for all, since many simple people—homemakers, fishermen, etc.—possess more intelligence (so that a minimal repertoire suffices to think assertively about experience and the world) than some lettered folk, especially the ideologized, who remain crystallized under a single sort of influence and set themselves up, foolishly, as supreme inspectors beyond good and evil.

The result is disastrous; by the fruits the tree is known. To keep this text palatable, instead of listing the evolution of problems and scandals—present to some degree in every country—I prefer to list the observed decline of love, joy, peace, patience, kindness, goodness, faithfulness, gentleness, and self-control.

It seems to me that the arrival of the new robotics, just around the corner, and cost reductions from scale production of humanoid (and other) robots will bring substitution of workers in repetitive tasks, thanks to a form of "hacking" of human motor tasks.

The Substantial Self: Zoom All

"After exhaustive hours of travel, he arrived just in time, on his way to his main destination; conversations quieted and the looks of the many family circles turned as he approached—not only due to the long wait for the last absent son, but also because of his resemblance to the main actor of that scene. After kissing his siblings and his mother, he saw him a few meters away, lying within a wooden hexagon, hands and mouth white. All that was transitory is now seen in the light of what is singularly definitive."

As for the protagonist of the novel *O Mestre Forjador Viajante* [*The Traveling Master-Smith*, unpublished manuscript, Joinville, 2025], his acts settled into a static schema. Any attempt to describe a human being is deficient, because it is an abstraction of a larger reality. Perhaps poetic discourse helps us more.

When noises—misunderstandings, incomprehensions, illnesses, and evils—fall silent, for those who possess forgiveness (from Medieval Latin *perdonāre*, formed by *per-* "completely" and *donāre* "to give, grant as a gift," from *dōnum* "present, gift"), the essence lets itself be known at a glance.

As in a musical score, we see all the notes, rhythms, and harmony at once; yet to know whether the music is pleasing, we need the passage of time. It is common ground that a single off-key note is unpleasant, and anyone can notice it in a performance, even if unable to explain what went wrong. Still, the whole cannot be judged by the isolated. Indeed, many pieces in minor keys are beautiful, even though they inhabit the so-called "sad key"; others, by containing dissonant notes, add quality to the harmony. It all depends on the whole—on the harmonic proposal. This is the symbolic view of the human being, of the substantial "I."

New Übermensch

Just around the next corner, we see a kind of "cavalry" arriving; integrated with artificial intelligence, it occupies the core of the issue: genetic engineering (CRISPR-based gene editing) + nanotechnology (interfaces/delivery systems), vectors frequently associated with transhumanist agendas. Experiments in this and allied lines have existed for years.

The German term *Übermensch* (super-man), coined by Nietzsche, names the human type capable of creating one's own values after the collapse of traditional ones since "God is dead." It is a critique of conformism: Nietzsche's "last man" (settled, without greatness) resembles the status quo some readings of transhumanism claim to overcome.

Untermensch (sub-man) was a racial and political label used by the Nazi regime to dehumanize and stratify peoples deemed "inferior" to "Aryans."

If the world's ecosystem is complex—for instance, consider the environmental impact that a simple change in a given bee species may cause—then altering the human being's natural constitution is a dangerous irreverence, even if the promise is tempting: to compete with robots or even to achieve immortality. If the reader is not a creationist, at least consider the anthropic principle: "There is a perfect adaptation of the structure of the entire universe to the physical-environmental conditions for human existence on Earth, such that if any change, however minimal, had occurred in the formation of galaxies, the human species could not have arisen." If there is no reverence for the One in whom all things were created and subsist, let there be reverence, at least, for the millions of years of evolution. For if man is the measure of all things, to see him as reconstructible is to fragment him—and to risk divorcing mind—Plato's *nous*—from living experience, from reality, confining oneself to a sort of fantasy limited to the most immediate physical perceptions. It is to risk losing sight of the articulation of an immense system of dynamics and latent possibilities. What we share is precisely how we are different.

Transhumanism in its eugenic currents is a modern rhetorical analogue to Nietzsche's *Übermensch*.

Suppose that, by agents of false science, those who refuse transhumanist biotechnological hybridization come to be hierarchized, would they be classified as "inferior," akin to the *Untermensch*?

Under the pretext of alleviating human suffering, are we not precipitously suspending the ethical limits of medicine and science, grounded in the implicit sovereignty of scientism?

Conclusion

Artificial intelligence, in itself, is neither bearer of benefits nor of harms; it will simply amplify human choices toward good or toward evil.

It is up to each human being, individually, to make the intelligent choice of the right path, as yet untrodden. By keeping technology at our fingertips, we can reclaim quality time for our families and remain present in the integrity of being.

Although the best of human essence has not evolved at the same pace as technology, I have no doubt that life will find a marvelous way and that the best is prepared for those who love goodness and truth.

About the Author

Christiano Gerhart Dittrich has a 25-year trajectory in industrial projects and ASME-compliant process equipment. A mechatronics engineer with a specialization in product engineering and emphasis on FEA, he is the founder of DittrichEng Ltda and, more recently, DittrichEng AI, a startup focused on the application of AI to engineering and to secure communication, including the TableHub.io platform. He has worked as a designer of critical jigs and fixtures for the Busscar NB1 production line, as well as on industrial projects related to the implementation of Caio Induscar's branch plant. A diligent investigation of his German roots resulted in dual citizenship and inspired the novel O Mestre Forjador Viajante, a work that explores themes such as technological development and resilience in the face of antagonistic situations. He endured several hours of captivity following an armed confrontation during a solo singing performance at a local church.

Throughout his career, he has accumulated achievements and lessons learned from his own mistakes, remaining deeply grateful to his parents, siblings, and the friends life has given him. He is married to a virtuous woman and is the father of Sofia—a cheerful and joyful girl, an avid reader who, at just seven years old, has already completed 490 books.

Email: christiano@dittricheng.com

Website: www.dittricheng.com

LinkedIn: https://www.linkedin.com/in/christiano-dittrich

CHAPTER 7

SWIPE RIGHT ON AI: EMBRACE AI AS YOUR NEW CAREER COACH

By Steven G. Edwards
Tech Entrepreneur and AI Enthusiast
West Palm Beach, Florida

The future belongs to those who learn more skills and combine them in creative ways.

—Robert Greene

Albert Einstein once suggested that intelligence is not about knowing everything, but about knowing how to find what you need when you need it. Now, here is the reality of today. AI is your assistant, not your enemy. But it cannot be your everything. AI can prepare you, but it cannot show up for you. It can sharpen your message, but it cannot replace accountability or effort. People who blindly rely on AI without thinking will fail just as fast as those who refuse to use it at all.

Two years ago, I used ChatGPT to write a love letter to my wife for Valentine's Day. We do not even celebrate the holiday, which made it even better. When she opened it, she started laughing, looked at me, and said, "You used your cheating tool to create this."

At the time, that is how a lot of people viewed AI. As cheating. As lazy. As cutting corners.

Fast forward to today, and that same "cheating tool" helps me every single day. Even better, my wife now uses it daily herself. The thing she once laughed at became something she relies on. Not because it replaced her thinking, but because it helped her move faster, think more clearly, and get unstuck when she needed help.

That is exactly how AI works in careers and in life. At first, people resist it. They joke about it. They dismiss it. Then quietly, they start using it. And once they do, they realize it was never about cheating. It was about capability.

AI did not make my message less meaningful. It helped me say what I already felt, but it was better. The same is true for job seekers, recruiters, and organizations. AI does not replace effort, experience, or character. It amplifies them.

You still must show up, think, and have to do the work. But now, you have a tool that helps you prepare, communicate, and compete at a level that was not possible before.

So here is the truth. AI is not a shortcut. It is a force multiplier. And just like that love letter, it does not make things less human. When used the right way, it actually helps you show up more human than ever.

The question is not whether AI belongs in your career. That decision has already been made for you. The only question left is whether you are willing to use it before everyone else does.

By the time you read this, you may already be behind, because AI is moving and changing daily whether you like it or not. That is not meant to scare you. It is meant to wake you up.

For decades, navigating the job market required money, access, or luck. If you wanted a strong resume, you paid for a resume writer. If you wanted interview prep, you hired a coach. If you wanted clarity on your career path, you hoped someone took the time to guide you. The system favored people who already had resources. That system is outdated.

Today, tools like Career.io and ChatGPT as well as other AI tools have flipped the script. What used to cost hundreds or thousands of dollars can now be done in minutes. Resume building, cover letters, interview prep, career exploration, salary research, job tracking are all available on demand, at scale, and backed by real data.

Fifteen years ago, I was running in-person job fairs. I coached people face to face on how to interview, how to research companies, how to dress, and how to follow up. I taught them how to track applications on spreadsheets because that was the best tool available at the time. I spent hours walking people through the basics because no one had ever shown them how hiring actually works.

One of the biggest lessons I taught was the difference between the sniper approach and the spaghetti approach.

The spaghetti approach is throwing the same resume at every job and hoping something sticks. Recruiters see it instantly. It wastes time and kills credibility.

The sniper approach is targeted, intentional, and disciplined. You tailor your resume to the role. You align your experience to the job description. You speak the employer's language.

Back then, tailoring a resume properly took hours. That was the excuse people used to avoid doing it. That excuse no longer exists.

With AI tools like Career.io's resume and cover letter builder, job seekers can tailor resumes to specific roles in under three minutes. Not three hours. Three minutes! The system translates experience into language applicant tracking systems understand, while still reading like a human wrote it. This dramatically increases the chances of getting past the ATS and into the hands of an actual recruiter. This is why organizations must change how they prepare people for work.

Colleges and universities cannot keep treating career services as optional. Military transition programs cannot rely on outdated workshops and generic handouts. Workforce development organizations cannot pretend AI is a passing trend. If you are responsible for preparing people for the job market and you are not teaching AI tools, you are failing them.

Career.io is one example of how AI can be used the right way.

For job seekers, the platform acts like a 24/7 career coach. Career insights show real labor market data, so people stop chasing jobs that no longer exist and start building skills that are actually in demand. Job seekers can see what industries are hiring, what skills matter, and what certifications move the needle before wasting time applying blindly.

The job search and job tracker tools bring order to chaos. Instead of sticky notes and spreadsheets, job seekers manage applications, interviews, and follow-ups in one place. This alone increases accountability and momentum, two things most job seekers struggle with.

Interview preparation is no longer about memorizing answers. AI-powered interview simulations allow job seekers to practice real scenarios and receive feedback on clarity, tone, and confidence. This is especially powerful for people who have not interviewed in years or are transitioning into civilian roles for the first time.

Salary tools remove guesswork. Job seekers walk into negotiations informed, not hopeful. They understand what the market pays and what their experience is worth, which prevents underemployment and frustration later.

Career pathways help people stop asking what job they should apply for and start asking what path makes sense for them. That mindset shift changes everything.

For transitioning soldiers, this impact is even bigger. Career. io helps soldiers translate military experience into civilian language, explore careers aligned with their skill sets, and build resumes that actually make sense to civilian employers. Transition Assistance Program (TAP) counselors can use the platform to guide soldiers at scale instead of relying only on one-on-one conversations that cannot keep up with demand.

The learning platform gives soldiers access to certifications and upskilling opportunities before separation. That means they are not just leaving the military with experience, but with credentials employers recognize.

For workforce development managers, Army TAP managers, and college counselors, the real power is analytics. They can see engagement, track job search activity, identify who is falling behind, and measure outcomes. This data matters. It supports funding requests, reporting, accreditation, and program improvement. It also allows staff to step in early instead of reacting when it is too late.

Case managers save significant time by automating repetitive tasks. That time gets reinvested into meaningful conversations, strategy, and accountability. This is not about replacing people. It is about letting people focus on what technology cannot do well.

Recruiters need to hear this too. Recruiters who resist AI will become irrelevant. The best recruiters already use AI to screen, source, and communicate faster. What separates great recruiters is not avoiding AI but using it to remove friction so they can focus on relationships, judgment, and decision-making.

AI is fundamental to your future. The job market has already moved on. The only question is whether you are willing to move with it. How you use it will help you or hurt you. That is your choice. You can ignore it and hope things go back to the way they were. Or you can swipe right on AI and embrace it as your new career coach.

About the Author

Steven G. Edwards is a workforce technology executive, author, entrepreneur, AI advocate, and TEDx speaker focused on how artificial intelligence is reshaping careers, recruiting, and leadership. He is the president of Premier Virtual by Career.io, a virtual hiring and career technology platform that has powered more than 10,000 virtual hiring events and supported over one million candidates and tens of thousands of employers across the globe.

A former member of the U.S. Army's 82nd Airborne Division, Steven brings a disciplined, no-nonsense approach to innovation and execution. He took Premier Virtual from idea to acquisition in under four years and continues to lead at the intersection of AI, recruiting, and career readiness. His work centers on helping organizations and individuals move faster, think smarter, and stay human in an AI-driven world.

Steven is the author of *Weeding Through the BS—From Idea to Acquisition* and is an instructor at Florida Atlantic University for the Veterans Florida Entrepreneurship Program, where he helps veterans turn ideas into scalable businesses while embracing modern technology and AI-powered tools. He is a frequent speaker for workforce development boards, colleges, military transition programs, and enterprise leaders, known for cutting through hype and focusing on real outcomes.

Above all, Steven is a husband and father who believes technology should not replace people, but empower them. His perspective on AI is grounded in real life, real work, and real responsibility. He challenges leaders and job seekers alike to adapt quickly, use AI intentionally, and remember that tools do not define success, people do.

Email: Steve.Edwards@career.io

LinkedIn: https://www.linkedin.com/in/stevenedwards23/

CHAPTER 8

DESIGNING MEANING: HOW JUDGMENT, NOT AUTOMATION, DECIDES WHAT MATTERS

By Russell Goldman
Founder, More Wow and Buildable Engine; AI Strategist
New York, New York

I had selected his features as beautiful. Beautiful!—Great God!
—Mary Shelley, *Frankenstein*

When the explorer John Lloyd Stephens pushed through the mosquito-laced undergrowth of the Yucatán in the early 1840s, he hoped he might be among the first outsiders to glimpse the architectural wonders built by the Maya. Storied but undocumented, what he discovered exceeded his greatest expectations. In his published account of this "discovery," he described his first view of the ruins of Uxmal as having driven "everything else from our minds … Our feelings were of wonder, admiration, and incredulity."

Obviously, Stephens was not the first person to see Uxmal. Modern Maya communities had lived among those ruins for generations. But for Stephens, the moment was a revelation. Standing among the vines and shadowed palms, what overwhelmed him was not simply the scale of the structures, but the unmistakable intentionality of a civilization that had chosen to construct monumental forms not to house offices or government services but as expressions of cosmology, identity, and belief. They were material manifestations of meaning.

Although it is easy to forget this architectural heritage, design has always been the first tool humans have used to give meaning to forms. Before language, industry, or technology, we had builders arranging objects into stories.

Now, as we enter the era of the fourth Industrial Revolution, led by accelerating advances in artificial intelligence, we should consider not only how AI can help us but also how we can use this technological wave to continue investigating what meaning means for us—and to infuse our world and our designs with what matters to us as individuals, communities, and cultures.

Technology Across Human History

This is not the first time humanity has been confronted with a moment like this. When humans first transitioned to settled farming around 12,000 years ago, they freed themselves from the daily labor of hunting and gathering and were liberated to pursue deeper callings. The world experienced an explosion of artisans, architects, priests, astronomers, healers, weavers, and storytellers. Leaving behind mechanical labor opened the door to new forms of identity, design, and collective culture.

AI is now beginning to free cognition itself. Although many of us are concerned that if left unchecked, AI may create enormous economic risk—or worse—I believe AI's deepest potential is in unlocking new forms of meaning for all of us. What can happen when we automate the technical bottlenecks that consume our creative energy? Will humans return to the work that has mattered most: art, imagination, and the crafting of our own destinies? That is my hope.

Design as a Process for Meaning

Design may begin with inspiration, but it comes to life through process. Across disciplines—architecture, interior design, industrial design, and digital product design—the process is surprisingly similar.

Good designers are great listeners and translators, investigating how imagination, materials, and constraints can create something new. While clients may expect an output that appears meaningful, for designers the adherence to process is itself a method for identifying meaning.

From there, designers map intentions and gather requirements to define what a space or product must do and what it must feel like. They identify the emotional center and narrative of a project and work to articulate the internal anchors that focus the work and reduce noise. When a designer tells you that a home must feel like a sanctuary, that a monument must feel like a national memory, or that a chair must feel like a hug—this is the work being done.

With the objective clarified, designers use a variety of methods to bring something to life. Sketches become diagrams, diagrams become layouts, and layouts become models. Designers test proportions, adjust light, revise materials, refine pathways, adapt interactions, and wrestle with the nuances that make form resonate. In architecture, interiors, and digital products alike, this might mean refining a floorplan, adjusting the flow of a mobile interface, or tuning the way light, materials, and interactions shape how something feels.

Meaning accumulates through thousands of decisions and emerges from an intentional balance of philosophy, necessity, and aesthetics. When a space or product tells a hidden story about the people who inhabit it, it becomes newly resonant as a union between what is built and what is felt. It is the choreography of relationships between people and space, objects and identity, tools and purpose.

As we move deeper into the age of AI, we will no doubt see a proliferation of design tools. Some will offer increased efficiency, while others may seek to replace designers at lower cost. For consumers and business owners to understand the trade-offs to come, it is important to understand design not only as art or as procedure, but also as the union of both: something integral to defining our sense of meaning in the world today.

Mexico as a Masterclass in Continuous Meaning

If there is a place where meaning and architecture visibly intertwine, it is Mexico. Like Stephens, I've always felt that there, the landscape itself feels intentional. Pyramids rise from corners of the country not as monuments to rulers but as extensions of the sky.

In Mesoamerica, as in other ancient cultures, design was not intended to be simply ornamental; it was a way to convey cosmic instruction. Builders understood the world as an interconnected system linking nature, humans, gods, and celestial forces. From pyramids to ball courts and temples, the architecture mirrored that worldview. The stones became repositories of belief reminding us that meaning is something humans can build. Even modernists understood the power of these lineages. Diego Rivera and Juan O'Gorman built Museo Anahuacalli from volcanic stone as an homage to the Mexican landscape and to the country's design heritage. Its sharp geometric archways and rugged materiality function as a modern shrine to cultural memory that deliberately bridges the past and present.

Architecture has always been an argument about identity. The impulse to infuse buildings with meaning is as clear in Mexico as it is in the neoclassical monuments in Washington DC. There, designers borrowed the visual language of the Greeks to root a new nation in an inherited philosophy of democracy.

Nowhere is the relationship between design and meaning more visible than in the world's sacred architecture. Temples, mosques, pyramids, cathedrals, and synagogues—these are humanity's original technology for shaping awe. From Borobudur to Gaudí's Sagrada

Família to Anahuacalli, these structures are unforgettable not because they offer shelter, but because they offer significance. Everything in these buildings is purpose built—from the ascent of light and the geometry of buttresses and thresholds to the calculation of acoustics and the way stone absorbs or reflects sound.

Stand inside the Sagrada Família as sunlight streams through stained glass and fractures into a million brilliant colors, you will feel your body become small inside a space designed to make you look up and recalibrate your existence. The parallels between a Mayan pyramid and a cathedral run deeper than their differences. Both are cosmological expressions and emotional instruments that rely on craftspeople whose knowledge of the design process has been carried across generations. To enter these buildings is to be reminded that design can alter both our surroundings and our internal states. Sacred architecture teaches us that meaning is engineered.

Our Crisis of Meaning

In design, the environments people inhabit have become increasingly void of feeling and story, often following aesthetics shaped by social media algorithms or by an internet-to-editorial echo chamber that rewards trends with short life spans. As a designer whose career partially depends on satisfying editorial requests, I can tell you that we are living in an era of infinite content and a scarcity of meaning.

I recently read an interview with the owner of one of the world's most respected architectural design firms who, when asked to debunk a design myth, replied: "You don't need a story for your [art] collection." Nothing could be further from the truth. When design role models and industry icons reject stories—the foundational element of meaning-making—we can be certain we are living through a crisis. The challenge is not a lack of beauty; it is a lack of intentionality. Perhaps artificial intelligence offers a turning point.

AI Neutrality

As many have pointed out, AI is a method of manipulating data to produce outcomes based on patterns. While the inputs and outputs hold meaning for us, the artificial intelligence itself is only a tool; it does not contain meaning. AI can generate possibilities and accelerate experimentation, but it remains fundamentally neutral. At its core, AI is a prism: It refracts the patterns it is fed into new permutations, but it does not interpret them.

In the field of design, the question is not whether AI will shape the future—it will—but whether it will deepen meaning or flatten it. Left unchecked, AI can create a world of hyper-generated sameness by producing infinite variations on the same visual language, or aesthetics that feel correct but hollow. But used intentionally, AI can free humans from the technical bottlenecks that prevent them from focusing on meaning.

Meaning emerges when humans have time, space, and freedom to think. AI might give that back to us. This is why I find this moment exhilarating: We may be gaining the possibility of more meaning.

Buildable Engine as a Case Study in Meaning-Centered AI

Earlier this year, my design firm, More Wow, created a tool that sits at the intersection of design and automation, a tool meant to return meaning to the architectural process by removing the weight of complexity. That tool, Buildable Engine (www.buildableengine. com), began with a simple question: Why does the process of creating anything—from the smallest object to a home to the largest commercial building—still require enormous manual effort, technical knowledge, and hours of repetitive work? Why does the path from idea to reality remain so inaccessible?

Architecture is filled with bottlenecks: code compliance, layout constraints, adjacency rules, accessibility requirements, technical annotations, zoning logic. All are necessary; all consume the time and energy designers need to think creatively. The challenge is not that architects and designers lack ideas; it is that the system around them forces them into the role of compliance technicians. Creativity, meaning, narrative, emotion—these are squeezed into whatever time remains.

Buildable Engine exists to change that. At its core, it is an AI system that carries the burden: a rules engine that understands code; a graph-based spatial-logic system that evaluates relationships; a fixing and healing engine that resolves conflicts; a layout intelligence that produces plans that are buildable rather than merely visually pleasing. It is a machine designed to free the designer to think, imagine, and create.

Originally, I also conceived of a tool that would generate plans end-to-end, including creative ideas. But architects consistently told me they had no problem with creativity; in fact, design is what drew them to the field. Of course. Creating a tool that enhances the source of meaning and reduces the time lost to tedium was clearly the more meaningful path. When Buildable Engine automates a compliance task, it is not replacing creativity; it is protecting it. It returns time. It reintroduces space for intention.

In this sense, Buildable Engine reflects a philosophical choice: the belief that AI should elevate human work rather than compress it. The belief that meaning in architecture and design comes from the human mind, not the machine. And the belief that the future of design will require systems that liberate us from the parts of the process that weigh us down. My hope is that this tool becomes an example of how AI can serve meaning rather than erase it.

The Good Scenario: AI as Meaning Multiplier

If used with intention, AI becomes a meaning amplifier. It can democratize access to design literacy, helping someone who never studied architecture to explore spatial relationships, test ideas, and understand constraints in ways previously unavailable. Instead of gatekeeping design, AI can expand its entry points and turn knowledge into something shareable and intuitive.

For designers, the benefits are even more profound. Rapid prototyping, for instance, when fueled by AI can allow for deeper exploration of concepts. Iterating to find the perfect option can become a form of discovery rather than a drain on time. You can now test ten ideas in the time it once took to test one. You can push the boundaries of your imagination and spend more time exploring how to make something resonate emotionally instead of wrestling with compliance.

When AI handles the mechanics, designers can return to the art, giving us more opportunities to ask deeper questions, like: What should this space feel like? What story should it tell? How should it change the person who enters or uses it?

This is the good scenario: a world where AI expands our capacity to express identity and intention.

In this future, I hope to see design filled with more narrative, architecture becoming more experiential, and human crafts gaining value. AI, in this future, does not flatten meaning. It magnifies it by giving us back the one thing we never have enough of: time.

The Bad Scenario: AI as a Meaning Destroyer

But there is another possibility, which is that AI can also accelerate the collapse of meaning.

We already see hints of this in the digital world. As mentioned earlier, we live in a time when the abundance of content has coincided with rising rates of depression and disconnection. If this mentality extends fully into the physical world, the loss will be far greater.

AI can destroy meaning not because it is malicious, but because it lacks intention. A machine cannot determine what is meaningful for the masses. If humans abdicate authorship, the machine will not replace it; it will simply generate what is most statistically probable. And what is most probable is rarely what is most provocative, intentional, or meaningful.

The risk is not that AI will design for us. There are situations where that may even be beneficial. The risk is that we stop designing altogether, and in doing so, sever the connection between art, story, and meaning. The bad scenario is a world where meaning becomes optional, inconvenient, and ultimately expendable.

New Human Skills: Curation, Translation, Narrative

This is why the future of design will belong to new kinds of practitioners, people who can curate, interpret, and translate. Designers will no longer be defined by their ability to produce drawings; machines like Buildable Engine will handle that. Instead, they will be defined by their ability to understand culture, encapsulate communal identities, craft emotional experiences, generate spatial ideas, and edit outputs that translate values into form.

The future designer will need more emotional intelligence, cultural literacy, and psychological insight to ask the questions that matter: Who are you? What do you believe? What do you want your life to feel like? How should your environment express your identity?

While machines can mimic creativity, uniqueness, and meaning with enough data, doing so at high fidelity may prove financially impractical. It's more likely that we will see a divide emerge with lower-cost, mass-market designs increasingly generated by AI and higher-end work that uses AI for process optimization but relies on human expertise to push creative boundaries.

This divide may seem inequitable. Why should thoughtful design be accessible only to those with greater means? Yet this distribution may also create a more equitable future. If AI enables the efficient development of apartment blocks, the overall cost of large-scale real estate projects could decrease, potentially making housing more affordable overall. This is my hope.

The Future of Architectural Meaning

When designers are liberated from the mechanics and when AI takes responsibility for the rules, constraints, and hidden labor that once consumed entire careers, we open the door to a new era of experiential space. Architecture could shift from being primarily a technical discipline to becoming, once again, an emotional and philosophical one.

Buildings may begin to respond to emotion; environments may shift with seasons, rituals, or the needs of their inhabitants. We may see hybrid structures that merge material craftsmanship with digital intelligence or sacred spaces that blend ancient symbolism with generative light and sound. Public buildings could be infused with stories the way ancient pyramids once were. Civic architecture could reflect collective identity rather than budget constraints.

This is where AI and humanity can form a true partnership. Machines can hold the boundaries, but humans should shape the meaning. Instead of optimizing for cost or speed alone, we could optimize for memory, awe, and connection. We could design not only spaces to inhabit but experiences that change how we inhabit ourselves.

Imagine a world where sacred architecture is not a relic of the past but an ongoing cultural practice—where the principles behind Chichén Itzá or Sagrada Família are translated into new forms that speak to contemporary life without losing the emotional gravity of ancient ones. AI can help us build this world; but it cannot and should not lead.

The future of architectural meaning will not be defined by what AI generates, but by what humans choose to ask of it. If we lead with curiosity, identity, and care, AI becomes a multiplier of meaning. If we abandon authorship, it becomes a generator of emptiness. The choice is ours, and that choice will shape the built world for generations.

Personal Reflection: More Wow and Designing a Life of Meaning

When I left my previous career as head of innovation at LG Electronics and entered the design world full-time, I wanted to realign my work with one core belief: Design is an emotional technology. I wanted to help clients tap into their own "wow" and create spaces and products that were transformatively meaningful. It is no surprise that my design practice is highly procedural—derived from software design—and closely tied to the process I outlined earlier in this chapter.

In the interiors side of my practice, we use technology to accelerate and enhance our outputs. We've developed AI tools that help us identify color palettes we hadn't considered and assist clients in identifying their stylistic likes and dislikes. Just as I hope Buildable Engine transforms the architecture, engineering, and construction industry, AI has also been transformative for More Wow. It has allowed me to remove noise from the process, enabling me to focus more on the human parts of design that have long been overshadowed by tasks and logistics.

AI has given me the time to cultivate the relationships I have with galleries and vendors, relationships that no machine can replicate. It has given me more space to create unique pieces for clients and to collaborate with artisans to infuse heritage-rich aesthetics into a space. It helps me create more wow with more meaning.

Designing Meaning in an AI World

We are entering a new era in which machines will handle an ever-growing share of our technical labor. But I believe that the role of the designer, the architect, the craftsperson, and the storyteller is not diminishing. It is expanding.

AI can accelerate work, but only humans can anchor that work in meaning. The future of design will belong to those who know how to translate between people and mediums and to those who understand that meaning is something we must build with intention.

Designers who embrace this will use AI not as a shortcut, but as a tool to deepen their craft and create work that is even more resonant.

Whether this future becomes real depends entirely on our willingness to reinforce meaning as a social value, on our willingness to build and use technology in ways that align with our values, and on our willingness to oversee, regulate, and recalibrate the systems we build. Whether liberating or confining, that future is entirely up to us.

About the Author

Russell Goldman is a multidisciplinary designer and former global AI executive whose work sits at the intersection of cultural history, emotional storytelling, and advanced technology. He is the founder of More Wow (www.morewow.design), a design studio focused on heritage-infused interiors that privilege feeling, identity, and craft, and the founder of Buildable Engine(www.buildableengine.com), an AI platform designed to transform creative intent into buildable reality.

Goldman previously led innovation and AI product strategy at LG Electronics, where he shaped global initiatives in generative and applied AI. He holds a BA from Princeton University and an MBA from The Wharton School and brings a global, interdisciplinary perspective to the future of design.

His work and perspectives have been featured in *Architectural Digest*, *Fast Company*, *The New York Times*, *Mansion Global*, *Southern Living*, and other major publications. Whether designing interiors, building technology, or writing about the future of craft, Goldman's work is unified by a single belief: that design is a form of emotional technology and that meaning is something we must build with intention.

Email: russ@buildableengine.com

Website: www.buildableengine.com

LinkedIn: https://www.linkedin.com/in/therussellgoldman

CHAPTER 9

POWER IN THE AGE OF DATA: HOW TECHNOLOGY AND AI REDEFINE GEOPOLITICS

By Robin Goudappel
Tech Business Leader, Keynote Speaker
The Randstad, Netherlands

A lie can travel halfway around the world while the truth is putting on its shoes.

—Mark Twain

For centuries, power in the international arena was measured through a simple set of tangible factors: the size of a country's territory, the number of its inhabitants, the strength of its economy and military, and its ability to provide energy, water and food for its people. Nations rich in resources—oil, metals, uranium, or fertile soil—were in a position to shape world affairs. Those that lacked them depended on others and, therefore, had less autonomy. In this traditional model, geography was destiny, and power politics was a game played mainly by states, using visible means of influence and enforcement.

In the last three decades, however, a new dimension has been added to this classical equation. Power is no longer just about land, people, and resources. It is increasingly about data, technology, cyber security, and artificial intelligence. Information has become both a resource and a weapon. The ability to collect, process, and control data determines economic advantage, military effectiveness, and social stability. The boundaries between physical and digital power have blurred, creating new kinds of dependencies and vulnerabilities. As one analyst put it, "Data is the new oil, but unlike oil, it can be copied infinitely and weaponized instantly."

This new dimension has fundamentally changed the way states and other actors exert influence. Data and technology touch every traditional element of power: They shape economies, enable military capabilities, influence populations through information flows, and determine how autonomous a country truly is in securing its own energy, food, and water systems. Artificial intelligence multiplies these effects by automating analysis, prediction, and decision-making. As a result, the geopolitics of the 21st century is no longer just a struggle over territory. It is a struggle over algorithms, access, and autonomy. Cyber security is safeguarding the confidentiality, integrity, and availability (CIA) of the data and information.

Traditional geopolitics assumed that states were the main actors in this struggle. Today, that assumption no longer holds. The number of relevant players has expanded dramatically. Alongside nations, giant private corporations, especially in the fields of energy and digital technology, have become geopolitical actors in their own right. Companies such as Google, Microsoft, Huawei, OpenAI, and SpaceX operate on a global scale that rivals or even exceeds the influence of many countries. Their products shape the daily lives of billions of people, their platforms mediate political discourse, and their research defines the technological frontier. They are the infrastructure on which modern societies depend, and in many ways, they are both the instruments and the subjects of power politics.

This shift also means that power has become more asymmetric and unpredictable. A small country can now play a disproportionately large role on the world stage if it commands a strong position in data, technology, cyber security or AI. Estonia, Israel, and Singapore, for example, have leveraged digital innovation to achieve global influence far beyond their size. Iran, Nigeria, and North Korea build their global power strategies around cyber security and technology capabilities. At the same time, dominance in the digital realm can be achieved relatively quickly compared to traditional power bases, through targeted investment, education, and innovation. That makes the world more dynamic but also more unstable. Sudden technological breakthroughs can upend established hierarchies. Cyber conflict can cross borders in seconds (not only bordering countries). And nations that once seemed peripheral can suddenly become central players in global security.

The integration of data and AI into geopolitics has also introduced entirely new forms of competition and conflict. Where once armies and missiles defined deterrence, today influence can be exerted through code. Cyber operations target not only critical infrastructure but also the integrity of information itself. Attacks on data systems and information can undermine trust, manipulate public opinion, or disrupt supply chains without a single shot being fired. The confidentiality, integrity, and availability of data—known in cybersecurity as the CIA triad—have become fundamental elements of national security.

This is where the geopolitics of information overlaps with the psychology of perception. The integrity of data is directly tied to the spread of misinformation and disinformation. Control over narratives, whether through social media manipulation, deepfakes, or algorithmic bias, can influence elections, weaken alliances, or destabilize societies. The United States and the European Union have accused Russia of conducting coordinated disinformation campaigns. China has been criticized for information control and influence operations in Africa and Southeast Asia. Western intelligence agencies themselves engage in similar tactics, justified as defensive counter-operations. Each actor sees its own actions as necessary for protecting sovereignty, proof that in the digital age, truth itself has become a contested domain of power.

Artificial intelligence amplifies both the potential and the peril of this new era. Modern AI systems can process enormous volumes of data, identify patterns invisible to humans, and automate complex decisions. But they are also vulnerable to a range of specific attacks. Data poisoning can corrupt the training sets on which AI models depend, causing them to learn false patterns. Adversarial attacks—tiny, invisible manipulations in input data—can make an AI misclassify images or signals. Model inversion and theft threaten intellectual property and privacy, while prompt injection in generative AI can make systems reveal confidential information, manipulate the results, or behave unpredictably. These are not theoretical risks; they have already been demonstrated in laboratories and, increasingly, in the wild.

Protecting data and models has, therefore, become a central concern. Governments and companies are investing heavily in data governance, encryption, and digital audit trails. Techniques such as differential privacy and federated learning aim to balance innovation with confidentiality, allowing AI models to learn from decentralized data without compromising individuals. The concept of model integrity—ensuring that algorithms are not tampered with during their lifecycle—has emerged as a new security priority. For nations, the challenge is strategic: how to secure their digital infrastructure without closing themselves off from global innovation.

AI security also extends to the operational realm. The supply chain of AI—from open-source code to deployed systems—creates new attack surfaces. Malicious dependencies or compromised APIs can be exploited to infiltrate models or manipulate outcomes. Monitoring for anomalies, ensuring model transparency, and integrating AI into incident response have become essential practices. In this context, the European Union's AI Act and the General Data Protection Regulation (GDPR) are not just legal frameworks. They are instruments of geopolitical power. By setting global standards for transparency, fairness, and accountability, the EU seeks to assert "digital autonomy," balancing innovation with the protection of citizens and values.

Ethical and governance challenges are equally important. AI systems can reproduce or amplify social bias, leading to discrimination and reputational damage. Ensuring fairness, explainability, and human oversight has become part of the broader conversation about responsible technology. The principle of the "human in the loop" reflects a deep political concern: the fear of losing control to autonomous systems. Whether in autonomous weapons, predictive policing or credit scoring, societies are debating where to draw the line between efficiency and accountability. The outcome of these debates will shape not only national laws but the global moral order of the digital age.

At the same time, AI is not only something to be protected. It is also becoming a crucial defensive tool. Machine learning is now used in threat detection, automated incident response, vulnerability management, and fraud prevention. AI can sift through massive log files to detect anomalies in real time or predict which vulnerabilities are most likely to be exploited. Cybersecurity companies increasingly rely on AI to manage the scale and speed of modern attacks. Yet this creates a paradox: As defenders automate, so do attackers. The same technologies used to protect systems can be used to penetrate them. In the next two years, experts expect AI-driven attacks to multiply in volume and sophistication, as offensive operations become faster, more adaptive, and more difficult to trace.

This dynamic resembles an arms race but one fought in code rather than steel. Every nation and major corporation now faces the same dilemma: to remain competitive and secure, they must invest in AI. But by doing so, they also become targets. The stronger a country's position in data and technology, the more attractive it becomes to cyber adversaries. Digital sovereignty, therefore, requires world-class cybersecurity and, ideally, a degree of technological autonomy. Many governments are now emphasizing the need for regional control over critical infrastructure, data centers, and cloud providers to avoid dependence on foreign powers. The European Union's push for a "trusted cloud," India's data localization policies, and the US CHIPS and Science Act all reflect this logic of technological self-reliance.

Yet absolute autonomy is impossible. The global digital ecosystem is too interconnected. Chips designed in the United States are manufactured in Taiwan using equipment from the Netherlands. AI models are trained on datasets that span the entire internet. Open-source software, while essential for innovation, also creates shared vulnerabilities. This interdependence complicates the notion of sovereignty: Security cannot be achieved through isolation but through resilience, collaboration, and transparency. The challenge is to build trust in a world where data flows ignore borders and where no single actor controls the full technological stack.

Different powers approach this challenge through their own lenses. The United States prioritizes innovation and market dynamism, trusting its private sector to maintain technological leadership. China pursues a model of state-driven digital control, integrating AI into governance and national strategy under the banner of "civil-military fusion."

Let's look at two contrasting examples:

The United States of America: The US focuses on innovation, competition, and private enterprise. The government generally plays a supporting role, setting broad rules but allowing the private sector (companies like Google, Microsoft, OpenAI, Amazon, etc.) to lead technological progress. The US trusts that free markets and entrepreneurship will keep it ahead in areas like AI, cloud computing, and cybersecurity. This approach reflects the American belief in market dynamism and minimal government interference.

People's Republic China: China takes the opposite approach. It has a state-driven model. The government directly steers and funds technology development, aligning it with national goals such as economic growth, surveillance capabilities, and military modernization. Its "civil-military fusion" means that technological advances in civilian sectors (e.g., universities, startups, telecoms) are intentionally shared with the military to strengthen national defense and geopolitical influence. This reflects China's emphasis on centralized control, coordination, and integration of all parts of society to serve strategic objectives.

The European Union positions itself as a normative power, promoting human-centric regulation and ethical standards. Smaller nations, from the Gulf states to Scandinavia, focus on specialization and agility, aiming to carve out niches in cybersecurity, quantum computing or AI ethics. None of these approaches is inherently right or wrong; each reflects a particular history, culture, and understanding of autonomy.

The interaction between these models defines today's geopolitical landscape. The US–China rivalry over AI and semiconductors is not merely economic. It is about who sets the rules of the digital world. The war in Ukraine has shown how cyber and information warfare are integral to modern conflict, with both sides using AI to analyze intelligence, control drones, and create and counter disinformation. African nations, long viewed as digital followers, are becoming battlegrounds for technological influence, with Chinese infrastructure investments competing against Western connectivity projects. Even private companies are forced to take geopolitical stances, deciding whether to comply with sanctions, protect user privacy, or align with national interests. The old distinction between state and corporation, public and private, domestic and global, is fading fast.

In this environment, the concept of "digital non-alignment" is emerging: a desire among many countries to benefit from global technology without falling into dependency on any single bloc. India, Brazil, and several Southeast Asian nations are experimenting with open-source AI frameworks and domestic cloud initiatives to maintain flexibility. Their goal is to preserve strategic autonomy in a world where data flows are increasingly politicized. This echoes the Cold War's non-aligned movement but with new tools and stakes: algorithms instead of armies, platforms instead of territories.

The path forward will require balancing openness with protection, competition with cooperation. A fully closed, nationalist approach to technology would stifle innovation and fragment the internet. But an entirely open, unregulated system would leave societies vulnerable to manipulation and exploitation. The ideal lies somewhere in between: a democratized model based on shared, transparent standards, complemented by secure, region-specific applications that respect local laws and values. In other words, an architecture that combines open-source foundations with responsible, closed-source implementations.

Whether the world can achieve such a balance remains uncertain. The temptation to weaponize technology for short-term advantage is strong. Yet the long-term stability of the global system depends on mutual restraint. Just as the nuclear age required new norms and treaties, the AI age will require a new understanding of digital power, one that recognizes interdependence as a fact, not a weakness. If nations can accept that each has a legitimate right to autonomy while acknowledging shared responsibility for the digital commons, a more stable equilibrium may emerge.

Power politics, in its essence, has not disappeared. It has simply moved into new domains. Territory still matters, but so do algorithms. Armies still deter, but so do networks. Influence now spreads through code, not just through diplomacy. The challenge for this generation is to ensure that the tools we build to enhance human capability do not become instruments of division. In the end, the struggle over data, technology, and AI is not only a contest of power. It is a test of our collective wisdom.

About the Author

Robin Goudappel is an international business leader with a strong vision for how technology can advance both society and security. With decades of experience in the tech and cybersecurity sectors, he combines strategic insight with a proven ability to translate ideas into effective execution. His expertise spans cybersecurity, AI, and digital transformation, helping organizations accelerate innovation and performance while staying aligned with customer needs. Educated at Nyenrode Business University and INSEAD, Robin has built and led successful teams worldwide, balancing entrepreneurship with a deep understanding of people and culture. In addition to his executive roles, he supports European startups through DutchBasecamp and YES!Delft and contributes to thought leadership on the geopolitics of technology. As an author and speaker, he focuses on the intersection of data, power, and digital autonomy, advocating for a secure and human-centered technological future.

Email: robin@alegriaventures.nl

CHAPTER 10

TEACHING CARS TO SEE: EYES THAT NEVER BLINK

By Sowmiya Narayanan Govindaraj
Machine Learning Engineer, Robotics
Los Angeles, California

The real voyage of discovery consists not in seeking new landscapes, but in having new eyes.

—Marcel Proust

A Moment That Changed How I Saw Machine Vision

There's a moment from my work that I return to often. It wasn't dramatic. It was simply the first time I sat in an autonomous vehicle as it began to move on its own. The steering wheel turned itself with a precision that felt both familiar and foreign, like watching someone write in your handwriting.

I remember looking around the cabin, noticing how ordinary everything felt, despite the extraordinary reality that the car was perceiving and interpreting the world without my input. It recognized traffic lights, tracked nearby vehicles, and responded to subtle changes in the scene, all with a calm attentiveness.

That first ride stayed with me because it clarified something no technical document had. Autonomy was not about complex models running underneath the surface. It was about supporting human capability and providing consistent attention when ours can falter. In that moment, machine perception felt less like science fiction and more like a straightforward extension of how we already see and respond to the world.

Why Teaching Cars to See Matters for Everyone

Most conversations about AI veer between abstract fears and distant futures. But autonomous vehicles address something immediate and urgent: human safety. Here's what we don't talk about enough. For generations, we've accepted a level of risk on our roads that we would never tolerate anywhere else in modern life. Over 90% of accidents stem from human error: distraction, fatigue, misjudgment, or simply being overwhelmed in the moment.

Level 4 autonomy offers a different path forward. But it doesn't run on algorithms alone. It begins with perception: the ability for a vehicle to understand its environment with clarity and consistency. When a vehicle truly "sees," mobility becomes safer for everyone sharing the road.

How Autonomous Vehicles See the World

If you sit in the passenger seat of an autonomous vehicle, it's easy to forget that the car is constantly paying attention. Unlike humans, it never blinks, never checks a notification, never looks away. Its "vision" comes not from one pair of eyes, but from a whole team of sensors working together.

Reading the World at a Glance

The vehicle's cameras function like human eyes, capturing colors, shapes, and details. They read traffic lights, recognize stop signs, and spot pedestrians crossing the street. Cameras help the vehicle identify what objects are and understand visual information that requires color and detail.

Feeling the World in 3D

LiDAR provides a different kind of "sight." It sends out thousands of laser pulses per second and measures how long they take to bounce back. This creates a precise 3D map of everything around the vehicle. While cameras show what things look like, LiDAR reveals where they actually are in space and how far away they are.

A Conversation of Sensors

Real power emerges from synthesis. A camera detects something moving. LiDAR pinpoints its exact location and trajectory. Radar measures its velocity. Each sensor validates and enriches the others' data, creating a composite understanding more robust than any single input stream could provide. Think of it less as one sophisticated instrument and more as an ensemble of specialists cross-checking their observations.

A Machine That Knows When It's Unsure

Perhaps most importantly, these systems are designed to recognize uncertainty. When the data is ambiguous or conflicting, autonomous vehicles slow down, create more space, or transition to fallback behaviors. They're built around caution, an attribute we don't usually associate with machines but perhaps should.

When all these layers come together, the result isn't artificial intelligence in the sense of science fiction. It's augmented perception:

machines supporting human safety through consistent, unwavering attention.

The Human Benefits of Machine Vision

The real impact of autonomous perception begins long before a vehicle hits the road. It starts with the people whose lives would change most dramatically. Safety improvements are immediate and concrete. Consider the everyday scenarios where human reflexes fall short: a child pursuing a soccer ball into traffic, a cyclist threading between cars, someone stepping off a curb while absorbed in their screen. These critical moments pit even the fastest human reactions against the limits of nerve conduction and muscle response. Machines maintaining continuous attention transform these situations from potential disasters into prevented incidents.

But safety represents only the beginning. For significant populations, autonomy means independence restored. For example, adults whose aging eyes or slowing reflexes have ended their driving years, but who still want engagement with their communities. People managing disabilities in a transportation infrastructure built around assumptions of standardized ability. Individuals are constrained not by what they can do, but by what their environments permit. These vehicles could shift mobility from something contingent on personal capability or economic access toward something approaching a universal right. Even for those who drive comfortably today, the benefits accumulate. Commuting hours reclaimed from white-knuckle navigation. Time redirected toward reading, conversation, rest, whatever matters more than gripping a steering wheel in traffic.

Environmental gains follow naturally. Vehicles that perceive precisely can optimize acceleration, eliminate wasteful idling, and select efficient routes. Better perception translates directly to reduced emissions and lower energy consumption across entire transportation networks. Machine vision offers more than technical progress. It provides leverage for addressing fundamental questions of safety, access, and sustainability in how we move through the world.

Navigating the Ethics of a Machine-Driven Roadway

Whenever technology steps close to human life, questions of ethics rise quickly, and rightfully so. People worry about accountability, about difficult decisions, about whether machines will be forced to solve impossible dilemmas on our streets. The actual design philosophy is more pragmatic than these thought experiments suggest. These vehicles are not programmed to choose between two harms in a split second. They are designed to avoid harm altogether. Everything—every algorithm, every redundancy, every validation test—is built on the principle of minimizing risk before it appears.

The architecture of autonomy is built on layers: sensing, prediction, planning, and control. Each layer has its own safety nets, and each decision is recorded in meticulous detail. Unlike human decisions, which vanish into memory, autonomous decisions leave a trail: traceable, reviewable, and improvable. That transparency alone shifts the ethical landscape in ways human driving never could. But ethics also means fairness. A perception system must see *everyone*, regardless of clothing, skin tone, movement patterns, or the quirks of lighting and weather. The obligation is not for the system to work *most of the time* or for *most people*, but for *all people in all conditions*. Safety cannot be selective.

Across the industry, teams spend countless hours testing and retesting these systems against bias and rare scenarios, from dimly lit alleyways to reflective vests, from snowstorms to city glare. Every failure becomes a lesson, every lesson a refinement. In a world where AI sometimes inspires fear, autonomous vehicles offer a more reassuring model: systems that are continuously audited, continuously improved, and held to a higher standard of accountability than any human driver could achieve alone. Technology does not replace ethical responsibility. It makes ethical responsibility even more essential. And when done right, it becomes a reflection of the principles we insist our technologies uphold: safety, fairness, and care.

Beyond the Fear: Understanding What Autonomous Vehicles Change (And What They Don't)

Fear is a natural response to the unknown. Autonomous vehicles ask people to rethink something deeply familiar: the act of driving. It's no surprise that questions emerge: Will technology eliminate jobs? Will the vehicle be watching me? Will machines someday make decisions that humans should? But the truth, shaped by years of building and deploying these systems, is more nuanced. Autonomy does not erase work, it evolves it. As vehicles learn to navigate on their own, humans take on new roles: overseeing fleets, maintaining systems, interpreting data, ensuring safety, coordinating logistics, and guiding the technology's ethical boundaries. These are jobs rooted in judgment, communication, and oversight; skills machines cannot match.

Autonomous systems are also far from being all powerful. They are bounded, tightly monitored, and designed with strict privacy and operational limits. Every mile driven comes with layers of human review and supervision. These vehicles are not independent intelligences roaming freely; they are deeply structured tools, shaped and guided by the humans who build them. What autonomy truly changes is not human importance, but human responsibility. The focus shifts from reacting behind the wheel to shaping how mobility works at a societal scale. Engineering teams pore over scenarios, testing teams study edge cases, operations teams ensure real-time oversight. Far from removing humans from the equation, autonomy increases the need for thoughtful, skilled, ethical human involvement.

Anxiety about replacement often obscures a key distinction: Machines won't assume roles requiring human strengths, i.e., navigating ambiguity, exercising empathy, generating creative solutions, providing care. They'll compensate for areas where we struggle—fatigue, distraction, reaction delays—while humans remain essential where we excel.

Thriving in a Future with Smart Machines

AI and robotics are becoming infrastructure rather than novelty. The people who adapt will find opportunities. Those who don't will face challenges. These tools have escaped the laboratory. They're showing up in education, small business, household management, and creative work. You don't need engineering credentials to participate in or benefit from this shift.

Learning to work with AI to enhance thinking, automate tedious tasks, and process information will become as fundamental as email literacy or smartphone competence. Yet as machines grow more capable, distinctly human skills become more valuable. Situations requiring empathy, nuanced judgment, cultural sensitivity, or ethical reasoning remain our domain. Communication ability, cross-disciplinary collaboration, emotional intelligence in complex contexts—these capacities will define success and leadership.

Most critically, this technological moment demands continuous learning. Innovation won't pause to let anyone catch up. All of us, regardless of age or field, need to maintain curiosity, flexibility, and willingness to grow. The most durable preparation isn't mastering any specific technology. It's cultivating the disposition to keep learning.

A Future Built on Shared Vision

When I think back to that first ride, I do not remember the technical details scrolling across my screen. I remember noticing how ordinary everything felt while something extraordinary was happening. The vehicle was perceiving the world with steady attention, filling gaps in awareness that humans naturally have. That quiet moment captured what teaching cars to see is really about. It is not the replacement of human drivers. It is the thoughtful extension of human capability.

As autonomous systems become more common, we won't simply find ourselves living in a world filled with smart machines; we will find ourselves shaping it. The choices we make about ethics, design, inclusivity, and imagination will determine whether

AI strengthens our humanity or challenges it. But if we build with intention, curiosity, and empathy, the path forward becomes clear: a safer, more accessible, more deeply connected future where machines do not replace our vision but expand it.

As our machines learn to see with clarity, they remind us to see with intention because while their vision maps the road, ours maps the future.

About the Author

Sowmiya Narayanan Govindaraj is a senior machine learning engineer specializing in perception systems for autonomous vehicles. His work spans multi-modal sensor fusion, 3D vision, efficient neural network architectures, and real-time deployment on automotive-grade hardware. He has contributed to next-generation robotaxi perception at Motional AD Inc and previously advanced Camera-LiDAR object detection and safety-critical deployment pipelines for autonomous trucking at Torc Robotics. Beyond engineering, Sowmiya's interests center on embodied AI and the future of human–robot collaboration, specifically, how intelligent machines can enhance safety, mobility, and accessibility in everyday life. He is passionate about designing systems that reflect human values, complement human judgment, and expand equitable access to autonomy. He also mentors aspiring roboticists and contributes to open-source and research initiatives aimed at bridging the gap between cutting-edge models and real-world deployment. He strongly believes the future of autonomy rests in its ability to extend human capability responsibly.

LinkedIn: https://www.linkedin.com/in/sowmiyanarayanan-g/

CHAPTER 11

FROM DATA TO DIALOGUE: DESIGNING HUMAN-CENTERED AGENTS IN THE AGE OF AI

By Jonathan Fraile Gutierrez
AI Leader and Product Builder
Dubai, United Arab Emirates

A Late Night in the Age of AI

On a Wednesday that had technically already become Thursday, Sofia was lying in bed, eyes burning, phone above her face.

The feed knew exactly what she clicked on: anxiety, burnout, high performer with no energy, nervous system hacks, sleep. It offered her therapists in 30-second clips, doctors explaining hormones, coaches talking about boundaries, creators doing so-called real talk about stress, and now, increasingly, smooth AI-generated advice.

She saved five posts in 10 minutes. She screenshotted a breathing technique. She wrote "Start this tomorrow" in her notes app for the twelfth time that month. And then, like most nights, she fell asleep without doing any of it.

If you zoom out, our age of AI disruption looks a lot like Sofia's night, just at civilization scale:

- We have never had so many experts visible: therapists, doctors, coaches, educators, creators.

- We have never had so much AI to amplify them: chatbots, copilots, agents, synthetic voices.

- And yet, individuals and organizations still struggle to turn good guidance into sustained, safe action.

The bottleneck is not insight. It is infrastructure.

In my work with KoAI Consulting, helping enterprises design AI architectures, and as co-founder of a self-care project that uses retrieval-augmented generation and voice agents, I see the same pattern everywhere. Most struggles are not just emotional problems or business problems; they are data and workflow problems.

This chapter is about how that realization led me from corporate AI strategy to designing small, safe agents that live much closer to the human nervous system and what that means for anyone trying to thrive in the age of AI.

When "I Just Need More Discipline" Is an Architecture Bug

When Sofia tells her friends about her stress, she says what most of us say: "I know what I should do. I just do not do it" or "If I were more disciplined, this would be fixed."

But if I diagram her situation like an architect, a different story appears. Her self-care life is a messy, ungoverned data system:

- Inputs: therapy sessions, occasional doctor visits, WhatsApp advice from friends, Instagram reels, YouTube talks, AI conversations, stray newsletter screenshots.

- Storage: notes apps, email, PDF attachments, bookmarks, saved posts, her own memory.

- Routing: whatever her tired brain happens to recall at 11:57 p.m.

- Interface: one overloaded human trying to be planner, database, and execution engine at the same time.

If this were an enterprise system, nobody in their right mind would ship it to production because:

- There are no defined sources of truth.

- There are no trust boundaries.

- There is no explainable graph of who to listen to for what.

- There is no event driven way to turn information into action.

The takeaway: Self-care, coaching, and even parts of healthcare are now fundamentally information systems. If we design those systems badly, people will suffer, no matter how strong their intent is.

AI does not change that by itself. If we just pour models on top of this chaos, we get what we already see: more content, more noise, more paragraphs that begin with "you should."

Where AI does help is when we start treating self-care and expertise delivery with the same rigor we apply to enterprise architecture:

- Who are the trusted sources?

- How is knowledge captured, normalized, and connected?

- What agents are allowed to act and under what constraints?

- How do we move from data to dialogue and from dialogue to small, grounded actions?

That is where my worlds began to merge: KoAI's work with AI in large organizations and a parallel experiment in building a self-care avatar that could sit between professionals and people like Sofia.

The Expertise Industry Is Fragmented. Humans Feel That Fragmentation

Before we ever talk about models, it is worth respecting the ecosystem that already exists, what I think of as the expertise industry. Behind all the content and programs, there are real people doing real, demanding work:

- Therapists and psychologists

- Doctors, nurses, home care teams

- Coaches, trainers, mentors

- Serious creators who have turned lived experience into structured guidance

In my consulting work, I often see these professionals at the edges of big systems: inside insurers, hospitals, wellness platforms, HR programs, and care technology companies. Their knowledge is deep. Their time is finite. Their content is scattered across platforms that were never designed to talk to each other.

Sofia does not experience this as architectural fragmentation. She experiences it as emotional whiplash:

- On Monday, her therapist tells her to track body signals.

- On Wednesday, a creator tells her to change her mindset.

- On Friday, a doctor's portal gives her lifestyle recommendations in dense medical language.

- On Sunday, she consults an AI that does not know any of this and gets a generic answer.

None of these people are wrong. But as a system, they are uncoordinated.

One of the core questions behind platforms like Kizuri is: What if we could build a place where people intentionally choose their experts, and where those experts' knowledge is structured to work together instead of competing for attention?

Imagine a marketplace where:

- You can find a therapist, a coach, a pain specialist, or a trusted creator.

- You subscribe not just to their feed, but to a structured, licensed version of their content.

- That content is clearly separated from random internet noise.

That is the first move: intentional selection of voices. But selection is not enough. We then need a way to connect what those experts say with what the person actually lives and feels. That is where the graph begins.

Building a Graph: Experts, Users, and the "Thin Slice" of Each Conversation

In KoAI projects, when we design AI around complex workflows such as insurance, home care, sales, or service, we rarely plug a model straight into raw logs. We:

- identify trusted sources

- normalize them

- define entities and relationships

- let agents act inside that graph

A similar pattern emerged when we started designing for self-care. In a self-care avatar, the graph is made up of three kinds of nodes:

- Expert content: the therapist's exercises, the doctor's non-diagnostic educational material, the coach's frameworks, the creator's long form explanations, all approved, attributed, and chunked into meaningful units.

- User provided information: the person's goals, history, and constraints; their own notes and reflections; any medical or contextual information they choose to share.
- Conversation extracts: the thin slice of each interaction with the system, not full transcripts, but structured summaries of topic, intensity, sentiment, and patterns.

This last part is key. Every time Sofia interacts with the agent, via text or voice, the system takes a small extract:

- "Sofia mentioned work meetings, tight chest, and 'I am failing' tone."
- "She completed a two-minute reset."
- "Language shifted slightly from hopeless to 'This is hard, but I tried'."

Those extracts are not interpreted randomly. They are run through patterns designed with experts in mental health, coaching, and care to determine:

- What overwhelm often sounds like in language
- What early improvement looks like
- Where red flags lie, such as risk or crisis themes
- How to distinguish tired from sliding into something more serious

The result is a self-care graph, a living, evolving structure that connects what experts say, what the user reports, and what actually happens between sessions.

On top of that graph, we can finally talk about agents.

Personal Retrieval: Grounded Agents Instead of Confident Improvisers

General purpose language models are incredible at sounding wise. They are less incredible at knowing when to stay quiet.

If Sofia types into a generic AI, "I feel like I am breaking down before every meeting," the model will cheerfully generate something based on everything it has seen on the internet. It does not know her heart rate, her medication, her therapist's strategy, or what worked for her last week. It just guesses, elegantly.

In the systems we build for enterprises with KoAI, and in experiments like Kizuri, we try to avoid that pattern. Instead, we use retrieval-augmented generation anchored in the self-care graph. The flow looks more like this:

1. Sofia speaks or writes: "I am on the edge again before this 4 p.m. meeting. Same tight chest."

2. The agent does not answer yet. It first retrieves from the graph what her therapist has said about early panic, how her coach framed high stakes situations, what Sofia did last time this pattern appeared, and what her current constraints are.

3. It composes a response based only on that retrieved material. The answer is not what the internet thinks; it is what her chosen experts and her own history suggest right now.

4. It labels and constrains its answer. For example, "This suggestion is based on your therapist's grounding technique and the micro steps you agreed on with your coach." If the detected pattern crosses into crisis territory, the agent stops and escalates: "This is beyond what I can safely handle; here are crisis options; please contact a specialist."

To Sofia, this does not feel like retrieval-augmented generation. It feels like a small, timely dialogue: "Last week, before a similar meeting, you tried a two-minute sensory reset and said it helped. Do you want to do the short version together now?" The model is not improvising. It is curating from a graph that binds expertise and lived experience.

This is one of the big contributions AI can make to care: not being smarter than humans, but being more consistent at bringing the right human wisdom back into the conversation at the right time.

Voice and Text: Agents That Fit into Real Life

"Nobody lives their life in a prompt box."

In enterprise projects, we have learned that the most successful agents are the ones that hide inside the tools people already use: CRM screens, inboxes, workflow apps, voice channels. Self-care is no different. The interface has to respect how people actually move through a day.

Sometimes Sofia is in a quiet room and wants to type. Sometimes she is in a car park, about to walk into a meeting, where typing is the last thing she can do. So the agent has to speak both text and voice.

Text is great for:

- End of day reflections

- Pattern summaries, such as "last month versus this month"

- Slower thinking

Voice is powerful for:

- In the moment micro support

- Eyes busy, hands busy situations

- Quick check-ins that feel less like using an app and more like being gently reminded

One afternoon, for example, Sofia agrees to a check in at 3:45 p.m. Her earbuds chime, and a calm synthetic voice, transparently labeled as AI but modeled with the consent and guidance of one of her experts, says: "Hi Sofia. You told me afternoons are rough. You have three minutes until your 4 p.m. meeting. Do you want to try the grounding exercise you liked last week or just skip today?"

If she says no, nothing bad happens. The system logs "skipped" and moves on. If she says yes, it guides her through a short, familiar sequence that she and her therapist agreed on.

Afterwards, it might add: "Noted. You completed a reset before the meeting. I will reflect this in your weekly summary, so you and your therapist can see the pattern."

That is it. No drama. No fake intimacy. Just small, repeatable support, acting as a connective tissue between sessions.

In terms of AI, this is where a lot of pieces meet: personal retrieval, expert designed patterns, light sentiment analysis, channel adaptation, and safety rules. In human terms, it is simply: "You do not have to remember alone."

Safety, Consent, and Scope: Patterns from the Enterprise World

When we design AI agents for large organizations at KoAI, especially in regulated industries such as insurance, healthcare, or financial services, three concerns dominate every architecture diagram:

- Identity: Who is this agent speaking as, and what is it allowed to claim?

- Consent: Who agreed to what, and can they see, control, and revoke that agreement?

- Scope: What is this agent explicitly not allowed to do?

The same questions are even more important when agents move closer to the nervous system.

If an agent is nudging you about your emotional state:

- It must always be clear that it is AI, not a human.

- It must not silently slide from support tool into unlicensed therapist.

- It must respect red lines: diagnosis, medication decisions, and crisis intervention that belongs to trained humans.

In practice, that means explicit consent flows for users and experts, clear labeling of synthetic voices, visible provenance for suggestions, such as "This comes from a specific program, adapted by AI," and robust crisis routing when certain patterns are detected.

I am convinced this is where enterprise and personal AI design will cross-pollinate heavily. The same patterns that protect a bank's customers or a hospital's patients can protect someone like Sofia when she talks to an agent at midnight.

For the AI world, this is a more demanding test than writing code or summarizing documents: Can we build agents that know when not to answer? Agents that say, "I do not know" or "I cannot safely help with this; here is how to reach a human," are not weak. They are trustworthy. In the long run, trust will matter more than raw capability.

What This Means for the AI Edge and for You

When people talk about the AI edge, they often mean faster execution, better automation, more leverage at work. All of that matters. In KoAI projects, we see teams win by building small, safe agents that reduce time to insight, time to action, and cost per task. But there is another, quieter edge: the ability to design agents that respect human limits and extend human care.

Sofia is not going to read model cards or inspect vector indices. She will simply live with whatever voices and suggestions show up when she feels worst.

You, however, as someone reading a book like this, can think about architecture:

- You can notice when a system is just dumping more data on you and when it is actually turning data into dialogue.

- You can ask where an agent's answers are coming from and whose expertise they are built on.

- You can choose tools that are honest about their scope and that embed real professionals, instead of pretending to replace them.

On the industry side, we have a responsibility:

- To blend enterprise-grade architecture with deeply human use cases

- To involve experts when designing patterns for insight and sentiment

- To route crisis and complexity to humans, not deeper into a model

- To see emotional infrastructure as just as real as any data pipeline

On the individual side, you do not need to build your own platform or design your own graph. You can start small:

- Decide whose voices you want near your nervous system

- Keep their best guidance somewhere you will actually see it

- Use AI as a connector between that guidance and your real day, not as a random oracle

- Track a single metric: how long it takes you, on a hard day, to take one kind action for yourself

In the end, civilization's next big disruption will not just be measured in model sizes or GPU clusters. It will be measured in how we manage the dialogues that surround us, the ones between humans and machines, and the ones inside our own minds.

For Sofia, the turning point was not a perfect app or a dramatic intervention. It was the slow construction of a better system around her:

- A few experts whose work truly helped

- A graph that remembered what she went through

- A small agent that nudged, listened, and knew when to stay in its lane

That is one version of the AI edge I want to see more of: not AI that overwhelms us with more brilliant data, but AI that quietly helps us move from data to dialogue and from dialogue to the small, human actions that actually change a life.

About the Author

Jonathan Fraile Gutierrez is an AI leader, product builder, and MarTech strategist with a background rooted in the Salesforce ecosystem and enterprise architecture. Over his career he has helped global brands design and implement data-driven customer experiences, bridging CRM, marketing automation, and analytics into cohesive, human-centered journeys. As founder of KoAI Consulting, he advises organizations on how to build responsible AI stacks, agentic architectures, and practical copilots that actually fit into daily work. He also co-founded Kizuri, a marketplace and self-care copilot that extends the reach of therapists, doctors, and coaches using retrieval-augmented generation and voice agents. Passionate about turning complex technology into usable products, Jonathan works at the intersection of strategy, architecture, and experimentation, helping teams move from AI slideware to real-world impact in marketing, sales, and customer care.

Email: jfraileg@koaiconsulting.ai

Website: www.koaiconsulting.ai

LinkedIn: https://www.linkedin.com/in/jonathanfrailegutierrez/

CHAPTER 12

LEARNING TO LEARN: THE NEW LITERACY FOR THE AI DISRUPTION

By Anat Heilper
Director of AI Architecture
New York, New York

The illiterate of the 21st century will not be those who cannot read and write, but those who cannot learn, unlearn, and relearn.

—Alvin Toffler

Let's be honest: The current conversation around AI is overwhelming. Even as someone who has spent over a decade building the very AI accelerator hardware that powers these new tools, I feel the speed. Every week, a new model is released, a new "impossible" feat is achieved, and a new wave of anxious headlines follows.

When I talk to leaders and professionals about AI, one word comes up more than any other: "fear." This is completely understandable. We are all feeling the ground shift beneath our feet. But that fear, and the anxiety of being "left behind" is not coming from the technology itself. It's coming from a "literacy gap."

When the printing press was invented, the power didn't just go to those who owned a press; it went to the small minority who knew how to *read*. We are at a similar inflection point, but this time, the new literacy isn't just about reading or writing. It's about learning. The real disruption isn't AI. It's the new, accelerated pace of learning that AI demands and *enables*. The challenge isn't "AI versus human." The challenge is, and always will be, "human *with* AI" versus "human *without* it."

If you don't use AI to accelerate your learning, you will be left behind.

That's a blunt statement, but it's also the most empowering truth of this new era. This technology, more than any other, is a tool for learning, augmentation, and growth. Your value doesn't disappear; it simply shifts. This chapter is your guide to making that shift, moving from a place of fear to a place of active, confident collaboration.

Part 1: The "AI Mindset" (Your New Foundation)

Before we can "work with" AI, we have to change how we *think* about it. In my two decades in the tech industry, leading teams of engineers and working with clients on new products, I've learned one thing: The best results never come from a perfect, all-knowing tool. They come from collaboration, iteration, and a healthy dose of skepticism.

I want you to stop thinking of AI as a magic box or an all-knowing oracle. I want you to start thinking of it as the most brilliant, tireless, and slightly clueless intern you will ever manage.

This "brilliant intern" has read virtually the entire internet. It can write code, draft a marketing plan, and summarize a 200-page medical journal in seconds. It never gets tired, never complains, and is eager to help.

But, and this is the critical part, it has zero real-world experience. It has no common sense, no "gut feeling," and no real understanding of consequences. It doesn't *know* what's true; it only knows what's *plausible*. It can "hallucinate" with incredible confidence, making up facts and sources that look completely legitimate.

A "passive user" takes what this intern says and "copy-pastes" it. This is not only lazy; it's incredibly dangerous.

An "active collaborator" (you) takes on the role of the senior manager. You guide the intern. You provide context. You check its work. You are responsible for the final product. This mindset transforms AI from a competitor into an "ally" that requires active management.

Part 2: The Practical "Learning Map"

Once you have the mindset, you need the skills. Thriving in this new era comes down to mastering three new (and very human) roles: the supervisor, the collaborator, and the strategist.

Role 1: The Supervisor (Your Critical Thinking Is the Value)

Your second key outcome—that critical thinking is a must—is the most important job of the AI supervisor. Your new added value is your judgment.

As I mentioned, AI models can and *do* make mistakes. They "hallucinate." They invent. They reflect the biases of the data they were trained on. The fastest way to destroy your credibility and add negative value is to blindly trust the AI's output. Your job as the Supervisor is to interrogate, validate, and contextualize:

- Interrogate and Validate: Never trust, always verify. When the AI gives you a statistic, a historical date, or a "fact," your first question must be, "Can I verify this?" A hallucination is just a confident-sounding error. Your critical thinking is the firewall against that error. If you're an engineer, treat the output like a piece of code that must be tested. If you're a writer, treat it like a source that must be fact-checked.

- Ask "What's Missing?": AI models are trained to be agreeable and to complete a task. They aren't trained to ask, "Is this a bad idea?" or "Have you considered the ethical implications?" or "What is the second-order consequence of this action?" That is *your* job.

- Inject Human Context: The AI has no "gut feeling." It doesn't know your company's culture, the subtle politics of an email, or the specific strategic goal you're *really* trying to achieve. Your value is in taking the 80% generic draft and infusing it with the 20% of wisdom and experience that makes it perfect.

This supervisory role is non-negotiable. It is the core of your new value.

Role 2: The Collaborator (Learning to "Speak AI")

This brings us to your third key outcome: learning *how* to work together. "Garbage in, garbage out" has never been more true. The quality of your AI-assisted work is a direct reflection of the quality of your "prompts."

A good manager doesn't just tell an intern, "Write a report." They provide a detailed brief. The same goes for AI. This is a new form of communication, and it's a skill you must learn.

- Provide Context and Constraints: Don't just ask a question; frame the problem.

  o Weak Prompt: "Write an email to the team."

  o Strong Prompt: "I am a tech leader. I need to write a concise, optimistic email to my team of eight engineers.

We just missed a deadline due to unexpected bugs, but we have a new three-sprint plan. The email should be empathetic but firm, and it must not assign blame. Write three options for me."

- Assign a Role: Tell the AI *who* to be. This is a powerful way to shape its response. "Act as a skeptical venture capitalist and critique this business plan." "Act as a patent lawyer and identify potential weaknesses in this invention description."

- Iterate: Your first prompt is rarely your best. It's a *conversation*, not a command. Talk to it. "That's a good start, but make it more concise." "Expand on point 2." "Now, write it for a non-technical audience."

- **Choose the Right Tool:** This is where your cost, time, and accuracy analysis comes in. Not all tasks require the most powerful (and expensive) AI model. You wouldn't use a sledgehammer to crack a nut. If you need to summarize a simple text for your own notes, a fast, "good enough" model is fine. If you are drafting a complex legal document or a sensitive piece of code, you must use the most accurate, high-fidelity model available. Knowing *which* tool to use is a critical strategic skill in itself.

Role 3: The Strategist (Your Vision is the Guide)

Once you've mastered supervising the AI (Role 1) and collaborating with it (Role 2), your job evolves again. You move from task execution to problem exploration. This is the most exciting part.

As a leader, my job has always been to look around corners, to anticipate problems, and to set a direction. AI is now the single greatest tool for doing this. Because AI can handle the "what is," you are freed up to focus on the "what if."

- Ask Bigger Questions: Instead of asking, "Can you analyze this spreadsheet?" you can now ask, "Based on the last five years of our sales data, what three customer segments are we overlooking?" or "Simulate the impact of a 10% rise in component costs on our supply chain."

- Innovate Faster: AI is a powerful brainstorming partner. You can use it to generate 100 different ideas for a new product, a marketing slogan, or a solution to a technical bug. As someone who holds patents in AI optimization, I can tell you that innovation is often about finding a new path. AI can help you map thousands of potential paths in the time it would take a human to map one.

- See New Connections: By feeding the AI data from different parts of your business (e.g., customer support tickets, sales figures, and manufacturing reports), you can ask it to find patterns and correlations that no human team would ever have the time to look for.

The strategist doesn't just use AI to do their old job faster. The strategist uses AI to discover what their new job is.

Part 3: Building Your "Learning Engine" (How to Stay Ahead)

This brings me back to my first, most urgent point: If you don't use AI to accelerate your learning, you will be left behind. The roles of supervisor, collaborator, and strategist are not static. They are constantly evolving, because the technology is constantly evolving. What worked six months ago is already becoming outdated. This isn't meant to scare you; it's meant to focus you.

You cannot "learn AI" once and be done. You must build a personal "learning engine," a sustainable system for learning how to learn with these new tools. It's overwhelming to face a tidal wave of new tools and updates. The key is not to try and learn everything, but to build a simple, repeatable habit.

1. Pick Your "Signal": You don't need to read every AI blog. Find three to five trusted sources—a newsletter, a specific expert on LinkedIn, a technical journal—and commit to reading them. Ignore the rest of the noise.

2. Have a "Sandbox": Create a low-stakes environment where you can "play" with new tools. This could be a personal project, like planning a vacation, or a small, non-critical task at work. This is how you build intuition. As a hands-on leader, I've always found that I don't truly understand a technology until I've tried to break it.

3. Dedicate "Practice" Time: Set aside 30 to 60 minutes every week to be a beginner. Use that time to try one new tool or one new prompting technique. This consistency is far more powerful than a frantic, all-day cram session once a year.

4. Unlearn and Relearn: This is the hardest part. The new "literacy" requires you to be willing to abandon old, comfortable workflows. You have to be willing to ask, "Is the way I've done this for 10 years still the best way?" If the answer is no, you have to be humble enough to "unlearn" that habit and build a new one.

This learning engine is your personal defense against irrelevance. It is the mechanism that turns the "threat" of AI into your single greatest professional asset.

Conclusion: Your New "AI Edge"

We started this chapter with the fear of being left behind. Let's end with the clarity of the path forward. The "AI Edge" is not a piece of software. It's not about becoming a coder or a machine learning expert. It is a fundamental shift in how you learn, how you think, and where you place your value.

The AI is your brilliant, tireless, and clueless intern. Your "AI Edge" is your willingness to be its manager. Your edge is the critical thinking you bring as a supervisor, the firewall that stops a confident-sounding hallucination from becoming a costly error. Your edge is the communication skill you build as a collaborator, the art of giving a precise, context-rich prompt that guides the AI from a generic draft to a brilliant final product. And your edge is the curiosity you deploy as a strategist, the vision to stop asking, "Can AI do my job?" and start asking, "What new, incredible things can we do together?"

The ground is shifting, but you are not powerless. You are the human in the loop. You are the source of the judgment, the ethics, and the strategy. AI is the tool. You are the learner. And in this new world, the learner is the one who will thrive.

About the Author

Anat Heilper is a technology leader with over 20 years of industry experience, including more than a decade in AI accelerator architecture. An expert in HW/SW/algorithm integration, she has a proven track record of taking AI concepts to product, collaborating with Fortune 500 clients on customized solutions. Anat is an innovator with multiple patents in AI optimization. She is a skilled strategist and team leader, having managed 80-plus engineers, who maintains a hands-on approach to new technologies.

LinkedIn: https://www.linkedin.com/in/anat-heilper/

CHAPTER 13

MANUAL MAGIC: TURNING CONFUSION TO CLARITY WITH AI—A PARABLE

By Penny Hopkinson
CEO, Manual Magic AI Limited, Author, Speaker
Warwick, England, United Kingdom

We are what we repeatedly do. Excellence, then, is not an act, but a habit.

—Will Durant, summarizing Aristotle

The Late Opening

The sandwich board outside Patisserie Pénélope read: "Open 8 a.m."

At 08:37, the café door was still locked. A queue had formed: commuters, a mum with a buggy, a builder in hi-vis tapping his watch.

Inside, Lila, 24, sharp, willing, stood with a paper binder in one hand and her phone in the other. The binder read "OPERATIONS

MANUAL" in bold capitals. Her phone displayed a group chat with 97 unread messages. She needed the new allergen drill, last week's price change, and the weekend rota. The binder contained the old drill. The chat was filled with opinions, not answers.

"Where's the update?" she asked.

Jamie, the duty manager, pointed to a shared drive on the back-office PC.

"Somewhere in there. Try 'FINAL_v6_new-new.'"

By the time they found the correct file, cross-checked a PDF, and texted a photo of a sticky note with the changed steps, the builder had gone to the chain across the road, and so had the mum. The queue gave up.

At 09:15, the door finally opened.

Mara, the operations director for the parent group, watched from the pavement. She didn't scold Lila or Jamie. She didn't need to; the empty till spoke loudly enough.

That evening, she wrote in her notebook:

We're not lacking effort, but we're lacking clarity. We are sitting on a mountain of scattered IP: documents, presentations, emails, videos, and knowledge confined to people's minds. The result is confusion, rework, and risk. I need a way to go from confusion to clarity. Fast.

The Postcard

A week later, Mara received a postcard from an old colleague: "Come to the Thursday Network, a quiet room above a bookshop. Ask for the Manual Wizard."

The bookshop smelled of paper and dust. Upstairs, a small circle of leaders compared bruises.

- Priya ran a chain of care homes.

- Ben looked after a field-service crew who kept supermarket cold rooms from failing.
- Sian was a partner in a mid-sized law firm.
- Tom was a deputy head at a sixth-form college.

They spoke in the shorthand of people who'd seen the same film.

"Version sprawl," said Priya.

"Missed handoff," said Ben.

"Conflicted guidance," said Sian.

"Staff read the policy after an incident," said Tom.

At the back of the room, a woman sat with a fountain pen and a calm expression. She'd heard it all before. The group called her the Manual Wizard. She didn't sell anything. She just listened.

When it was Mara's turn, she shared the café story. Then she asked: "How do I make the manual live where the work happens?"

The Manual Wizard leaned forward. "Stop chasing people around the content. Bring the content to the point of need. Short prompts, short checklists, short videos with captions. And a simple rule: AI helps; humans decide."

"How?" asked Mara.

"With a small plan that doesn't break the week," said the Manual Wizard. "Run an IP Consolidation Sprint. One day. Get the lot on one table, digital and physical. Then build your first working slice with Manual Magic AI® doing the heavy lifting. Humans keep judgment, standards and culture."

The group went quiet. Someone scribbled the phrase, "AI helps, humans decide."

The Sprint

Mara booked a room with a long table, a big screen, and good coffee. On the door, she taped a sign:

Clear the Table Day

Phones on silent. Opinions welcome. Evidence better.

At 09:00, the team arrived: Lila from the café, Jamie, the manager, Carla from HR, Michel, the food safety lead, and Ravi from IT, carrying a portable scanner. A trolley groaned under the weight of binders. A tray holding a USB stick contained "FINAL" files.

Mara set the tone. "This is not about blame. It's about finding what we've already created and putting it to work."

They piled up everything: documents, PDFs, SOPs, training slides, supplier notes, WhatsApp screenshots, and scribbles on napkins.

Ravi fed paper into the scanner. Michel read the latest allergen drill aloud. Lila acted out the opening routine as if she were on shift.

On the screen, Manual Magic AI® began to ingest. It didn't guess at standards. It sorted, clustered, and drafted structures. It took three versions of *How to Close* and proposed a consolidated checklist. It turned a 17-page PDF on handwashing into an eight-step guide with a 20-second clip list for a captioned micro-video.

"AI as helper, not hero," said Mara. "It proposes, we decide."

By lunch, the confusion looked less like chaos. A simple taxonomy appeared:

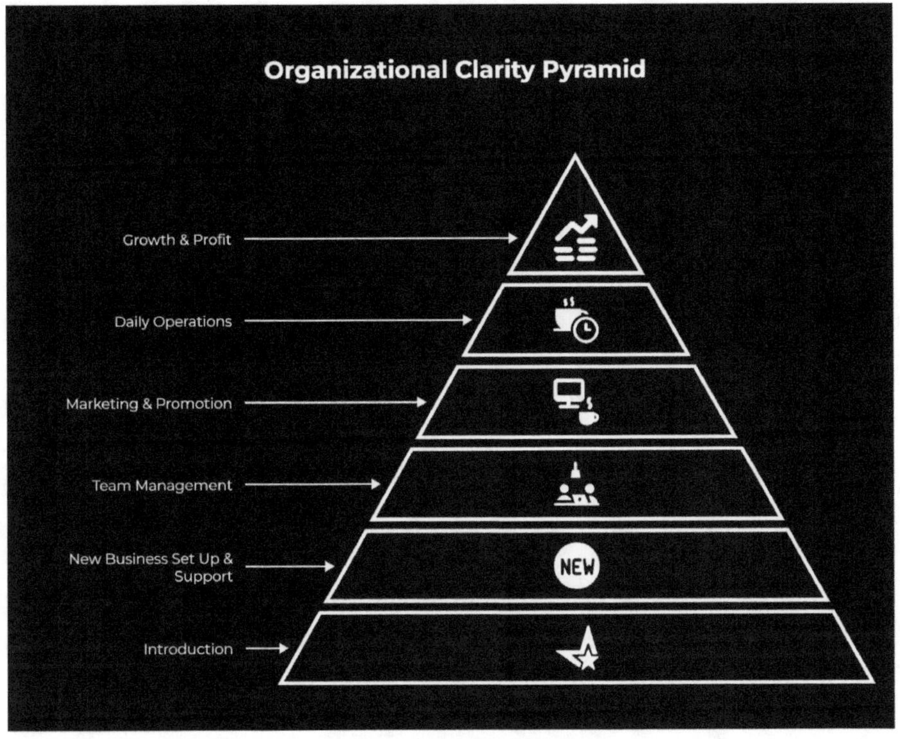

1. INTRODUCTION
2. NEW BUSINESS SET UP & SUPPORT
3. YOUR TEAM
4. MARKETING & PROMOTION
5. DAY-TO-DAY OPERATING REQUIREMENTS
6. DEVELOPMENT, GROWTH & PROFIT

They mapped the café world into those shelves. Not perfect. Good enough to move.

In the afternoon, they selected three priority SOPs: opening the café, allergen drill, and price change at the till.

Manual Magic AI® turned the long text into crisp steps and drafted a storyboard card for each: camera angle, what to show, what to say, and timecodes for captions. Lila volunteered to be the hands in the video. Ravi grabbed his cell phone and a tripod.

By 16:30, they had:

- One structured hub with search.
- Three draft SOPs with checklists.
- One storyboard and test footage for a 90-second video.

They wrote a promise on the whiteboard: "One team, one source of truth."

The First Pilot

They launched the pilot at a single location. A QR code was placed near the till and on the allergen binder. Staff could scan it, watch a short, captioned clip in their preferred language, and follow the steps without leaving their station.

Three agreements kept everyone honest:

1. **Guardrail words.** If a task could impact safety or compliance, the SOP was marked with a red tag. Two people had to approve it.

2. **Update schedule:** a monthly review of all red-tag items and a quarterly sweep of the rest.

3. **Attribution.** When a team member improved a step, their name was added to the changelog. Small rewards highlighted the effort.

The first week wasn't glamorous. Someone complained that the captions moved too quickly. Another thought the QR code was positioned too low. The team adjusted the code, slowed the captions, and re-shot one clip. Nothing heroic. Just care.

On Friday, the café opened at 08:00, and the builder returned.

The Five-Seat Table

The Thursday Network met again. This time, the Manual Wizard seated the leaders at a round table.

"Bring one story from your world," she said. "Show us your gains and your risks. Keep it real."

Retail and Hospitality—Mara's Update

Mara reported, "We focused on three priorities: opening on time, allergen safety, and price changes. We streamlined the steps and shifted the content to the job via QR codes and dynamic search. New staff can now follow the process easily without searching."

"What's at risk?" asked the Manual Wizard.

"Allowing the content to decay," Mara explained, "So, we created a calendar. Red-tagged tasks receive a monthly review."

Healthcare and Care—Priya's Update

Priya shared, "In care, clarity saves dignity. We turned a five-page handover into a two-minute checklist, accompanied by a short clip. Captions by default."

"And the risk?"

"Treating the tool as an oracle. We keep a two-person sign-off for any step that touches medication or consent forms," Priya explained.

Field Services—Ben's Update

Ben shared, "We replaced a 22-step paper form with a straightforward flow, with each step visible on the phone. Photos are attached to the job card. Less back-and-forth."

"And your risk?"

Ben answered, "Signal in car parks. We enable the key flows to operate offline and then sync them afterwards."

Professional Services—Sian's Update

Sean told everyone, "We once believed manuals were only for kitchens and warehouses. We were wrong. Client intake, conflict checks, QA on deliverables—these are our 'how-tos.' We used Manual Magic AI® to turn long 'lessons learned' memos into checklists and templates."

Education and Training—Tom's update

Tom said, "Teachers drown in policy PDFs. We pulled the top ten 'moments of need': safeguarding steps, lab setup, and trip consent packs. Micro-lessons under five minutes. QR by the lab sink."

The Manual Wizard nodded as each story was received. The table seemed less like five separate sectors and more like a single pattern: brief, clear, delivered at the moment of need; AI provides support; humans make the decisions; and the rhythm of review continues.

The Sceptic in the Corner

At the back, a man sat with folded arms. He'd kept quiet. Now he spoke, "This feels slick. But I've seen tools boomerang. First-month buzz, then drift. What stops this from turning back into a dusty binder, just more attractive?"

No one flinched. The Manual Wizard waved him in. "Name?"

"Colin. I run distribution centers. I'm allergic to silver bullets."

"Good," she said. "So am I. Here's the dull truth that keeps this going."

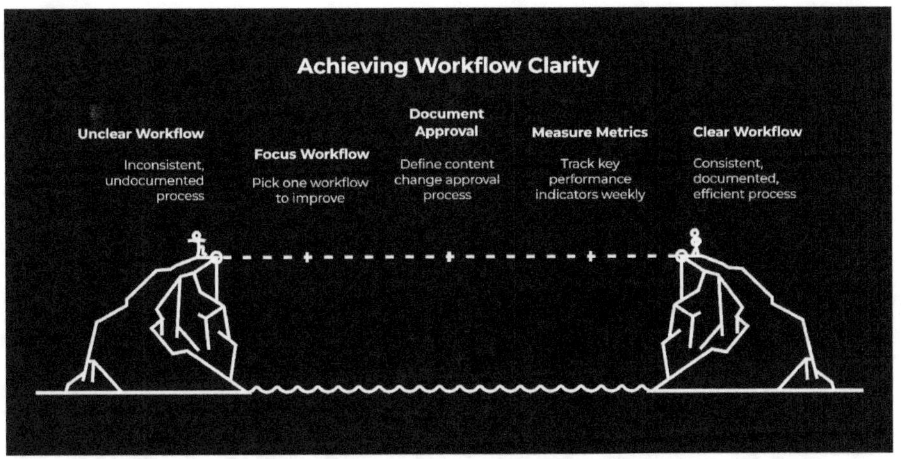

- Pick one workflow that matters. Not ten. One.
- Write down how content changes get approved. Stick to it.
- Measure three things weekly. You choose which.
- Put a face to each update. No anonymous edits.
- Make each video shorter than a coffee break.
- Book a slot every month to prune old content.
- Reward the people who fix friction.

Colin nodded. "That I can live with."

The 30-Day Postcard

Mara requested something she could give to her team. The Manual Wizard took out her fountain pen and wrote a postcard in capital letters:

30 DAYS TO A WORKING SLICE

Week 1 — Choose a single workflow. Run a one-day IP Consolidation Sprint. Decide your approval path.

Week 2 — Map to the six shelves. Draft 3 to 5 SOPs. Storyboard one video.

Week 3 — Pilot in one site or team. Gather feedback. Tighten roles and handoffs.

Week 4 — Measure, fix, and roll to a second site. Book a monthly review. Put the next three SOPs in the queue.

"That's it?" asked Mara.

"That's it," said the Manual Wizard. "Plain, not grand."

The Human Touch

A fortnight later, the café recorded the 90-second allergen drill on a cell phone. Lila was in the shot. She spoke clearly. Subtitles appeared beneath her hands. Jamie stood off camera with a checklist. Michel approved the wording. Carla added a line on respectful language. Ravi tested the QR code near the prep station.

The clip wasn't glossy. The lighting was decent, not cinematic. The steps were straightforward. Staff watched it once, then checked the oven.

A notice went up by the staff fridge:

"Shout out to Lila for making the allergen drill easier to use. $50 voucher on the way.

Next up: Deep clean in under 20 minutes. Volunteers?"

People smiled. Not because of the voucher, though it helped. Because the small piece of work felt like theirs.

The Boardroom Question

At month's end, Mara presented to the board. She didn't bring fireworks. She brought three figures, two photos, and a story.

- Opened on time: five days out of five at the pilot site
- Allergen drill: no misses in spot checks
- Price change: done in under 10 minutes each week

The photos showed a QR code near the till and Lila holding a tripod.

Her story was brief. "We were not short of effort; we were short of clarity. We consolidated our scattered IP into a single location, used Manual Magic AI® to organize and draft, and let our team establish the standard. We maintained a monthly rhythm. It may seem modest, but it works."

The finance director posed the familiar question finance directors ask, "What's the pay-off?"

Mara didn't guess. "I don't have a precise dollar figure yet. I have fewer delays, fewer back-and-forths taking up valuable time, which could be better used helping team members who struggle—and better mornings. We'll track the money over a quarter. For now, I'm asking to add two more sites."

The board agreed. Not because Mara dazzled them, but because she sounded like someone steering, not gambling.

The Evening Walk

On her way home, Mara stopped outside the café. The sandwich board said, "Open 8 a.m.," and, for the first time in months, it felt like a promise the team could keep.

She messaged the group chat: "Thank you. This begins to look like clarity."

A reply pinged from the Manual Wizard: "Clarity is a habit. See you next month."

The Plain-English Map

At the following Thursday Network, Colin asked the Manual Wizard to state the obvious—intentionally.

"Where does AI help?" he said. "Plain English. No fluff."

She obliged.

What AI Does:

- Organizes the mess. Clusters near-duplicates, detects overlap, and surfaces the latest versions.

- Initial outline formats. Convert lengthy paragraphs into bullet points and checklists for easier review.

- Condenses and converts. Summarizes a policy into a pre-shift reminder or a training deck into a job aid.

- Storyboards. Create a shot list and captions for short, watchable clips.

- Video. Converts videos into manageable steps, allowing users to watch, listen, or read at their own pace.

- Language. Translates speech into captions and converts it into the chosen language.

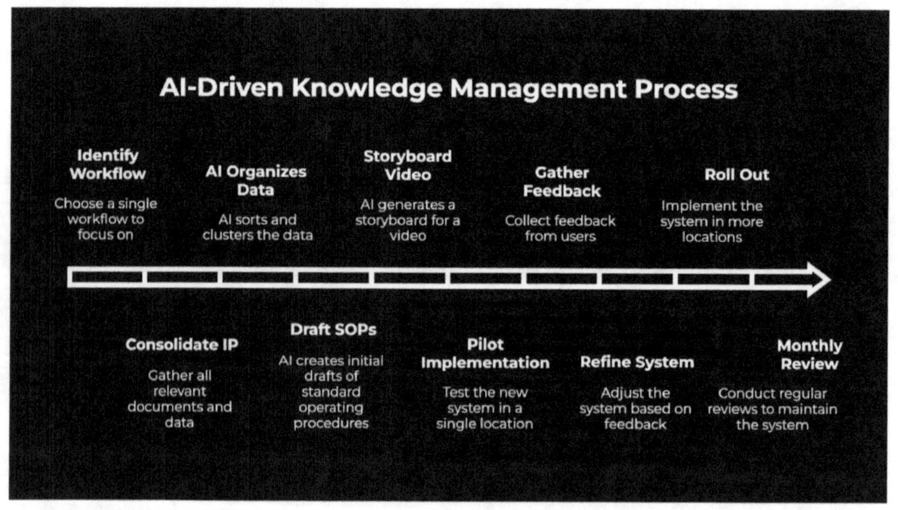

What AI Doesn't Do:

- Judgment calls. They can't be responsible for safety decisions or complex exceptions.
- Edge cases. It will miss the odd thing that happens once a year during a power cut.
- Context. It doesn't realize your site has a broken shelf unless you inform it.

"So, the line stays," she said. "AI helps; humans decide."

No one argued. They'd seen both sides with their own eyes.

The Quiet Governance Loop

Priya inquired about a way to keep the wheels turning without establishing a full-time bureaucracy.

The Manual Wizard drew a loop on the whiteboard:

Propose → Review → Approve → Publish → Use → Measure → Prune → Propose

Beneath it, she added six quiet rules:

- One inbox for proposed changes.
- Named reviewers for each category.
- Time limits on reviews: days, not weeks.
- Changelogs with names and dates.
- Red tags for safety-critical items with dual sign-off.
- A monthly pruning hour on everyone's calendar.

"It's not glamorous," she said. "It's what stops drift."

The Five-Generation Table

Tom raised a hand. "My staff spans five generations. Some love video. Some want a sheet of paper. How do I serve them all without building two systems?"

The Manual Wizard redirected the question to the group.

"Short videos with captions," said Mara. "Always with a printable one-pager linked underneath."

"Mobile-first access," said Ben. "But the one-pager is pinned in the workshop too."

"QR at point of need," said Sian. "Don't ask people to go on safari through a portal."

"Micro-lessons under five minutes," added Tom.

No one needed a seminar. They needed practical, human touches.

The Pitfall Parade

Colin requested a list of traps. The group complied.

The Traps:

- Creating videos that are too long.
- Treating search and naming as an afterthought.
- Skipping human review "just this once."
- Allowing each site to operate independently without a shared core.
- Gathering metrics that no one pays attention to.
- Expecting a tool to fix culture.

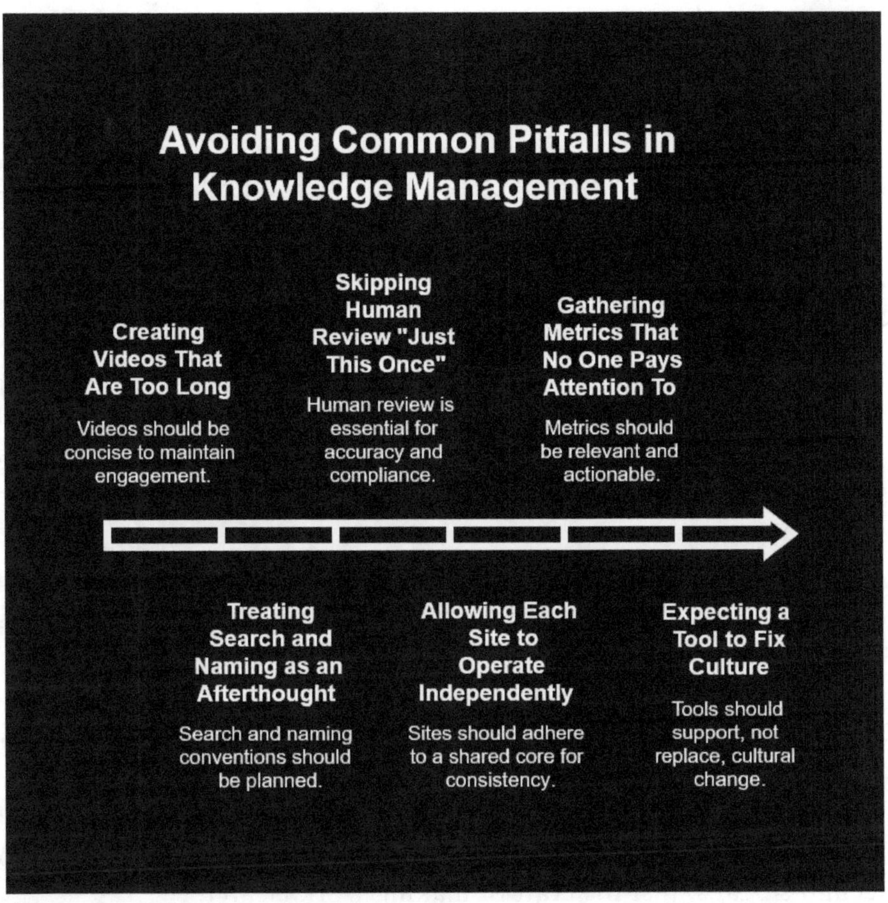

The Fixes:

- Keep videos shorter than two minutes unless there's a valid reason.

- Agree on a naming convention and adhere to it.

- Allocate dedicated time for reviews.

- Hold a thin core with layered local notes on top.

- Track three key numbers relevant to the work.

- Model the behavior: Leaders openly follow the same SOPs.

Colin nodded. "Not glamorous. But I can see how this holds."

The Second Month

Two months later, Mara observed something subtle. New starters shadowed for fewer days before they contributed. Seniors spent less time answering the same questions and more time coaching. The WhatsApp chat still buzzed, but it was gossip and shifts, not "Where's the latest drill?"

She said it aloud at the Thursday Network: "We haven't changed who we are. We've shifted where our knowledge resides and how we develop it together. Manual Magic AI® handles the sorting and initial drafts. We uphold our standards and judgment. That's the split that works."

The Manual Wizard smiled, then grew serious. "Keep the monthly pruning. Keep the red-tag checks. Keep the attribution. This isn't a project. It's a habit."

The Publisher's Bench

One afternoon, the bookshop hosted a small panel. An editor perched on a stool, listening as leaders shared their stories. No hype. No techno-slogans. Just the craft of making work clearer.

The editor asked Mara for a closing thought she could print without an asterisk.

Mara examined the room and kept it straightforward: "Clarity is not a speech. It's a series of small, dull, respectful acts—done by people who care about getting tomorrow right. AI helps us do the dull bits faster. People set the standard. That's how we move from confusion to clarity and stay there."

The room didn't clap. It exhaled.

The editor wrote the line down and underlined it twice.

About the Author

Penny Hopkinson combines decades of expertise in systems and operations with a passion for clear communication. Her career spans journalism, editing, and international reporting, equipping her to help organizations—from start-ups to global brands—navigate the challenges of creating robust operations manuals. As the visionary behind Manual Magic AI®, Penny has developed an AI-driven app that makes building standard operating procedures simple and effective, tailored to diverse learning styles and needs. Her acclaimed Manual Magic three-step system, detailed in her bestselling book, *Manual Magic—Create the Operations Manual Your Franchisees Need to Succeed*, continues to set the standard for operational excellence.

Learn more at: www.manualmagic.ai, www.pennyhopkinson.com, and www.linktr.ee/pennyhopkinson.

CHAPTER 14

THE HUMAN-LED AI ENTERPRISE

By Sara Jetta
Founder & CEO, JettAI Consulting
Toronto, Canada

Ultimately, it's not going to be about man versus machine. It's going to be about man with machine.

—Satya Nadella

How Leaders Steer Transformation in an Age of Intelligent Technology

The Whisper That Revealed the Truth

The moment that crystallized everything I believe about AI, leadership, and transformation didn't happen in a strategy workshop, an executive offsite, or an innovation lab. It happened behind a closed door.

A CEO, confident, composed, and normally unshakeable, waited until his team had left the room. Only then did he lean in, elbows on the table, and confess quietly: "Sara … it feels like everyone is building their own company inside my company."

There was no anger in his voice. No theatrics. Just honesty.

He continued: "Every team has their own tech pilots, their own tools, their own vendors. Nothing connects. I don't know what's real and what's noise anymore. And if I'm being honest … I don't know which decisions are grounded in fact and which are fiction."

It wasn't a technology problem. It was an identity problem. If a CEO no longer knows what's real, it isn't because the algorithms are too advanced; it's because the organization has become too fragmented for any one human to hold the story together.

That moment forced me to articulate something I had been observing for years: AI doesn't expose technical gaps; AI exposes human ones.

And the organizations that will thrive in this era are the ones whose leaders understand this fundamental truth: AI accelerates clarity when leaders are aligned and accelerates chaos when they are not.

This chapter is about that truth and about the human frameworks leaders need to succeed in a world where technology evolves faster than culture, behaviour, or governance ever can. It is also about something I care deeply about: keeping humans in the loop, not just as "approvers" at the end of a process, but as the authors of the vision, stewards of the data, and guardians of the impact.

The Human-Led AI Success Framework

The Five Pillars Every Modern Leader Must Master

This framework is not theoretical. It was built from years spent inside transformation programs, war rooms, boardrooms, and candid C-suite conversations across multiple industries. It's the pattern that reveals itself when you've been close enough, for long enough, to see what repeatedly works and what consistently fails. These are the five pillars every modern leader must master:

1. Alignment: Shared vision, shared priorities, shared language. Without alignment, AI amplifies fragmentation.

2. Data Coherence: Not more data, but meaningful, trusted, unified data. Coherence unlocks insight; chaos produces fiction.

3. Strategic Governance: Light where possible, strong where necessary. Governance is not control; it is protection, clarity, and empowerment.

4. Culture and Behaviour: Trust, communication, emotional intelligence. AI adoption lives or dies on the human side.

5. Execution Discipline (Prioritization and Scale): Fewer pilots. More outcomes. Organizations succeed when they choose focus over noise.

Each of the following sections brings a pillar to life through stories I've lived as a transformation executive, as an innovation leader, and as a C-suite advisor guiding organizations through complexity, uncertainty, and meaningful change.

Pillar One: Alignment

The Cost of Many Visions, One Company

I was once brought into an organization celebrated externally as an "innovation powerhouse." The branding was impressive: glossy labs, tech partnerships, bold mission statements about the future. The inside told a different story.

On my first day, a leader handed me a spreadsheet titled "Innovation Pipeline." Rows upon rows, scroll after scroll: thousands of pilots, POCs, algorithms, and experiments.

I asked a simple question, "Which of these reflect your enterprise priorities?"

Silence.

Finally, someone said, "We've never evaluated them collectively."

Another added, "We were told to innovate, so ... everyone did."

This wasn't innovation. This was organizational entropy disguised as progress.

A senior executive later confided, "Our pilots are multiplying faster than our progress."

That line has stayed with me because it captures the central risk of unmanaged AI enthusiasm: Enthusiasm without alignment creates fragmentation, not innovation.

And fragmentation is expensive: duplicated spending, inconsistent customer experiences, competing vendor ecosystems, incompatible data structures, leaders making decisions in isolation, and models that can't scale because they were never designed together.

I once walked into a meeting where three separate teams unknowingly piloted different AI chat agents for the same customer problem. All were proud of their work; none knew the others existed. Smart, committed people doing good work that would end up competing instead of compounding. AI didn't cause this. Lack of alignment did.

When we finally brought the teams together—product, operations, customer care, data, and finance—we didn't start with technology. We started with a whiteboard and one question:

"What problem are we actually trying to solve?"

It took an afternoon of hard conversations and humility. By the end, we had one shared problem statement, one outcome narrative, and one roadmap that multiple teams owned together. The AI agent we eventually launched was far better than any of the original three, not because the model was more sophisticated, but because the humans were finally working from the same story.

Leader Takeaway

If your teams cannot articulate:

- What matters
- Why it matters
- How success is measured

… then you do not have innovation. You have motion.

Human in the loop starts here: in the uncomfortable conversations where leaders choose one vision over many competing ones.

Pillar Two: Data Coherence

The Day I Discovered "Tens of Thousands of Reports"

One of the most revealing moments of my career came when I was leading a data and analytics transformation inside a complex enterprise. The symptoms were everywhere:

- declining trust
- contradictory dashboards
- leaders escalating decisions
- analysts burned out
- business teams building their own spreadsheets out of frustration

Everyone was swimming in data but starving for insight. Then came the real discovery: not hundreds of reports. Not thousands. Tens of thousands of active reports across the enterprise. Some were created years earlier and never retired. Some contradicted each other. Some counted the same metric in several different ways.

A VP looked at me and said: "I can get three different answers to the same question, all from official reports."

I remember saying something I've repeated at conferences since: "Many organizations are data rich but information poor."

The problem wasn't technology or talent. It was incoherence.

The risk became painfully clear when we learned that one of the legacy data warehouses feeding these reports was about to be decommissioned. Hidden in that environment were years of historical data certain teams depended on to understand seasonality, customer patterns, and risk exposure.

Had we simply "switched off" the old platform, we would have blinded parts of the business overnight. Decision-makers would still have their favourite reports. They just wouldn't know the underlying data had quietly disappeared. AI models trained on incomplete history would have confidently projected the wrong future.

So, we rebuilt the D&A function with the human at the center:

- We co-created consistent business definitions with domain experts.

- We rationalized the reporting ecosystem through hands-on workshops.

- We moved from "any report" to "trusted sources" with clear owners.

- We embedded business partners into each function to interpret results.

- We elevated intake governance so requests were tied to decisions.

- We restored a single source of truth.

Within months, leaders were telling me: "For the first time in years, I know exactly where to go for answers."

What they really meant was: "I trust the data again, and I trust the people behind it."

Trust is the currency of AI. Without it, no model will ever be adopted, no matter how brilliant. Human in the loop isn't just about approving an AI output; it's about humans designing the very definitions of truth that AI learns from.

Leader Takeaway

If your data isn't coherent, your AI will produce fiction with confidence and your people will eventually stop listening.

Pillar Three: Strategic Governance

The Enterprise Operating System for Intelligent Organizations

One of the most powerful lessons I learned about AI came from a transformation that had nothing to do with algorithms, at least not initially. I was asked to evaluate why a large enterprise felt constantly overwhelmed but rarely moved forward. Here's what I found:

- Thousands of active initiatives
- Nine separate governance committees overlapping, redundant, conflicting
- Leaders attending meeting after meeting
- No visibility into enterprise priorities
- Competing business cases
- No taxonomy for decisions
- No portfolio view
- No strategic focus

One executive finally told me: "We're not drowning in work. We're drowning in fragmentation."

He was right. This wasn't poor strategy. This was the collapse of an enterprise operating system, the structure every AI transformation eventually rests on.

So, we rebuilt that system from the inside out:

- We consolidated nine committees into four strategic portfolios.
- We aligned each portfolio to a clear enterprise priority.
- We assigned executive sponsors with real accountability.
- We empowered directors and VPs with decision rights.
- We integrated financial transparency and benefits realization.
- We introduced consistent prioritization criteria.
- We elevated visibility to the board.

The transformation didn't slow innovation. It created room for it.

At the same time, a large-scale system program had gone live and was causing downstream data issues and operational pain. Instead of creating a new "crisis committee," we folded stabilization into the same governance model. We created a cross-functional war room: architects, engineers, finance, operations, and business leaders under the umbrella of the portfolio council.

The message was simple: This isn't "IT's problem." This is our enterprise problem. Humans from across the system had to sit together, look at the same facts, and make hard trade-offs in real time.

It also taught me this: AI cannot scale without organizational scaffolding. AI needs governance, structure, and sequenced decision-making. Without it, AI accelerates chaos.

One of the most common misconceptions I encounter is the belief that AI reduces the need for governance. "The system will learn," people say.

In reality, AI increases the need for governance. Because intelligent systems reflect and amplify the operating environment they're built in. Human-in-the-loop governance means humans decide what's acceptable, what's ethical, what's worth funding, and when to press pause.

Leader Takeaway

Governance is not control. Governance is clarity, empowerment, and protection, the foundation AI needs to scale responsibly, with humans firmly in charge of the compass.

Pillar Four: Culture and Behaviour

The Hidden Operating System of AI

If I had to choose the single biggest determinant of AI success across every industry I've supported, it wouldn't be data quality, funding, or technical architecture. It would be culture.

I once walked into a company where three major groups were effectively at war:

- Merchants said data scientists "didn't understand customers."
- Analysts said merchants "ignored the data."
- Operations said both groups were "disconnected from reality."

The models weren't failing. The people were.

So, I brought together leaders from each group for a workshop. No slides. No complex language. Just conversation.

A store leader described the pressure of managing inventory during peak season. A data scientist explained how a small inconsistency in data can cascade through a model. A merchant shared the real-time stress of pricing under margin pressure while social media reactions played out in seconds.

People laughed, vented, debated, and eventually recognized that each function's pressures were real and interconnected. For the first time, some of them heard not just what the other group did, but what it felt like to be in their role.

Within weeks, adoption soared. Not because the model improved. But because trust improved. This is the true barrier to AI: fear, mistrust, and misalignment. Not algorithms.

Leader Takeaway

If teams don't trust each other, they won't trust the model, no matter how accurate it is. Human in the loop means inviting people into the journey early, not asking them to bless a decision at the end.

Pillar Five: Execution Discipline

From Pilot Paralysis to Scalable Impact

Across industries, I see the same pattern: Pilot theatre. Pilots everywhere. Movement everywhere. Progress nowhere.

In one organization, nearly 50 pilots were running simultaneously. All exciting. None scaling.

When I asked, "Which ones matter?" the room was silent. A few leaders tried to defend their favourites. Others genuinely didn't know what was happening outside their own silo.

So, I implemented a principle I still use today: No new pilots until three existing ones scale.

Suddenly:

- Leaders had to prioritize.
- Business sponsors had to engage.
- Teams had to collaborate.
- Vendors had to be held accountable.
- Impact had to be measured.

Innovation stopped being entertainment. It became purpose.

The biggest myth about AI is that experimenting widely increases your odds of success.

The truth is: AI success is not about volume. It's about focus, discipline, and the courage to stop.

Leader Takeaway

Pilots are not progress. Scaling is. Human in the loop means humans decide what deserves to scale based not only on what works in the lab, but on what works in the business, in the field, and in real lives.

Behavioural Economics in Action: The Digital Pricing Story

A major retailer once brought me in to understand why a digital pricing and convenience platform, one I had successfully deployed elsewhere, was hurting revenue in their environment.

The platform was performing exactly as designed:

- Real-time visibility
- Rapid product location
- Instant price updates
- Enhanced convenience

But an unexpected behaviour emerged: Customers found items too quickly. They spent less time browsing. Impulse purchases evaporated. Basket size dropped.

The issue wasn't the technology. It was human behaviour.

Convenience changed shopping patterns in ways leadership hadn't anticipated. What they saw as "frictionless" felt, in practice, like "get in, get out," which is great for speed, not always great for margin.

Turning it off would have been the easy answer. Instead, we got curious. We observed stores, talked to customers and associates, and looked at journey data, not just transactions. Then we rebuilt the strategy:

- We aligned features to business outcomes, not just novelty.

- We validated hypotheses using behavioural data and real observation.

- We implemented governance before activating new capabilities.

- We measured human impact before scaling.

We made deliberate choices about where to lean into convenience and where to preserve a bit of "good friction," the kind that allows for discovery, inspiration, and higher-value baskets.

This experience reinforced something I say in every AI leadership keynote: "AI doesn't operate in a vacuum. It operates in human psychology." Technology is predictable. Humans are not.

The Pattern Across Every Industry

Across every organization I have supported—retail, logistics, financial services, travel, QSR, consumer goods, and advisory—the pattern is universal:

- Leaders fear losing control, not to AI, but to fragmentation.

- Data coherence matters more than data volume.

- Behaviour determines AI success.
- Governance is the infrastructure of scale.
- Innovation requires disciplined prioritization.
- Culture drives adoption.
- Alignment is the real competitive advantage.

The Human-Led AI Success Framework exists because I kept seeing the same behaviours—the same failures, the same breakthroughs—repeated across industries, geographies, and organizational sizes. These aren't abstract theories. They're lessons learned in the trenches, with real people, real stakes, and real trade-offs.

Conclusion: Technology Accelerates What You Already Are

AI will not save an organization from itself. It will magnify whatever already exists. If you are aligned, AI will amplify that alignment. If you are fragmented, AI will accelerate chaos. If you are disciplined, AI will accelerate execution. If you are inconsistent, AI will accelerate confusion.

This is the truth leaders must confront: AI does not transform organizations. People do.

The leaders who thrive in this era will be the ones who:

- embrace alignment over speed
- choose coherence over complexity
- build trust across functions
- understand behaviour before models
- govern intelligently
- prioritize relentlessly
- scale with purpose

And above all, they will keep humans in the loop, not as a safeguard of last resort, but as the designers of the vision, the stewards of the data, the owners of the trade-offs, and the storytellers of what kind of future they are building. Because in an age of intelligent technology, the most important intelligence remains human.

About the Author

Sara Jetta is a senior transformation executive recognized for redefining how modern organizations think, operate, and lead in an age of intelligent technology. With more than 15 years of experience across retail, logistics, financial services, transportation, and consumer sectors, she has built a reputation for taking broken, fragmented systems and transforming them into aligned, high-trust environments where AI can scale with purpose and integrity.

Sara has built data and analytics organizations, launched AI and automation functions, architected enterprise governance models, and led transformation offices that directly influence C-suite strategy and enterprise-wide decision-making. Executives seek her out for her rare combination of strategic sharpness, technical fluency, and deep emotional intelligence, a blend that allows her to unify leaders, rebuild trust, and accelerate results in environments where alignment once felt impossible.

A respected keynote speaker and advisor, Sara brings honesty, clarity, and humanity to conversations often overwhelmed by hype. Through her advisory practice, she partners with executives to design the structures, behaviours, and leadership mindsets required to drive meaningful, sustainable transformation, always grounded in the belief that technology succeeds only when people do.

JettAI Consulting: strategy.speed.intellegent outcomes. "Where human-centered leadership meets intelligent technology."

Email: sara@jettai.ca

LinkedIn: www.linkedin.com/in/sarajetta

Website: www.jettAI.ca

CHAPTER 15

THE INTELLIGENT ENTERPRISE: HARNESSING THE AI EDGE FOR GLOBAL SCALE

By Olga Martynov
Financial Executive and Sustainable AI Advocate
Irvine, California

> *Man's mind is his basic tool of survival.*
> —Ayn Rand

The Emergent Intelligence: Civilization's Inflection Point

For the first time in history, humanity is building a rival to its own mind. Earlier revolutions transformed what we could do, but not what we could think. The steam engine amplified muscle, electricity expanded reach, the internet connected global knowledge. None of these inventions carried the capacity for learning or self-improvement.

Artificial intelligence is different. It learns, reasons, and accelerates beyond human comprehension if not guided wisely. As Yuval Harari warns, the greatest risk is not malevolence but misalignment, delegating decisions to systems we do not fully understand. MIT physicist Max Tegmark puts it bluntly: Intelligence is the most powerful force in the universe, and we are now creating more of it.

Yet Jensen Huang offers a contrasting best-case scenario: AI becomes more capable without drifting from human purpose. In this view, intelligence compounds inside boundaries we set, remaining interoperable with our systems, values, and workflows. Instead of the divergence some fear, AI quietly embeds into life as a continuously learning, aligned, and amplifying force.

These are not predictions of doom; they are reminders that the future depends on how we shape and align the intelligence we unleash. AI must remain transparent, interpretable, and anchored in long-term human purpose. Once intelligence becomes a shared global substrate, it must serve humanity rather than constrain it.

The economic stakes reflect this magnitude. McKinsey analysts estimate that AI-driven automation and augmentation could unlock trillions of dollars in annual value by 2030. To understand AI's role, we need to see it not as a tool but as a parallel, machine-generated form of cognition. We are building a cognitive grid, a shared intelligence substrate that everything will eventually plug into. Just as electricity powered machinery, this grid will power perception, reasoning, and coordinated action across people, enterprises, and societies.

AI's potential is expansive because its application surface is universal. Intelligence will no longer live only inside screens or software. It will increasingly diffuse into the systems that run economies and the infrastructures that support daily life, increasingly operating in the background rather than at the edges.

This is the defining power of the AI edge: the ability to make a billion micro-decisions faster and more accurately than competitors and to turn those advantages into durable differentiation. Competitive divergence will come less from any single model or dataset and more from how quickly enterprises learn, adapt, and integrate intelligence into their operating fabric.

We have seen something like this before. The digital economy democratized opportunity and lifted billions. AI is the next leap, not just connecting people, but amplifying their capabilities. When aligned with human values, it becomes a force multiplier for innovation, resilience, and long-term prosperity.

Long-term thinkers often ask: What must we build today so humanity thrives 200 years from now? AI makes that question urgent. Breakthroughs ahead will come less from scaling models and more from the fusion of machine intelligence with human insight.

Scaling AI safely requires more than algorithms, it needs reliable energy, sufficient compute, strong connectivity, and governance that protects human agency. In my work across finance and enterprise governance, one pattern is constant: Technology becomes transformative only when the infrastructure beneath it is accessible, trusted, and purpose-built for scale.

At its core, AI is a civilizational tool: an extension of human capability. When supported by trusted infrastructure and guided by enduring human values, it becomes not a source of disruption but a partner in shaping a thriving future for generations to come.

When Intelligence Becomes the Decision Substrate

For organizations, this civilizational shift, where AI becomes a pervasive layer of reasoning, carries immediate implications for how decisions are made and governed. Alignment is becoming the new fiduciary duty, and the responsibilities of leaders transform with it. It's on leadership to ensure that the intelligence shaping their organization stays true to its purpose, ethics, long-term value, and within the enterprise's intent. This is not a technology project; it is an organizational redesign of how judgment and accountability flow.

To understand this shift, consider how technology has evolved. For generations, technology expanded what we could do. AI now extends our cognitive capability. Its significance lies not in whether AI "thinks," but in the fact that intelligence is becoming the environment within which decisions occur. A tool can be controlled. An intelligent environment reshapes how an organization behaves. As AI becomes embedded everywhere, organizations won't simply use intelligence, they will operate inside it.

The parallel with past industrial inflection points is helpful but incomplete. Factories reorganized when electricity became abundant because movement could finally scale. Today's enterprises must reorganize around intelligence because judgment can now scale. AI changes how decisions are sequenced, how risk is perceived, how work flows, and how human and machine contributions fuse into continuous, adaptive systems.

At the Baron Capital Conference, Elon Musk described a 100-gigawatt orbital solar array powering AI, an image meant to highlight a practical truth: Intelligence scales only as far as the infrastructure beneath it. These constraints are economic and societal, not merely technical.

As intelligence permeates operations, the enterprise begins to resemble a living system: adaptive, predictive, and self-correcting. Efficiency becomes the shallowest outcome. The deeper value lies in adaptability: learning faster than competitors, anticipating threats earlier than regulators, and innovating beyond the limits of individual cognition. In this environment, the orchestration between human judgment and machine reasoning becomes central. AI may surface patterns, but leaders determine meaning. AI may optimize options, but boards evaluate risk. AI may accelerate outcomes, but finance ensures they align with fiduciary responsibility.

Autonomy introduces uncertainty. Systems that learn from massive datasets and surface insights no human explicitly encoded can strain traditional oversight. The key question shifts from "How do we adopt AI?" to "How do we ensure it behaves in alignment with our intent?"

Trust frameworks aren't red tape; they're the structure that makes AI safe to scale. In an AI-native enterprise, trust is not inspected after deployment; it is engineered in by design. Governance, compliance, and ethical guardrails must be woven into every step of the intelligence lifecycle so that systems behave reliably and securely. All of this signals a broader truth: Once intelligence becomes part of the operating fabric, governance must evolve as quickly as capability.

To succeed, organizations must think long term, not transactionally. That means updating governance and risk practices, so they keep pace with AI's role in decision-making. Leadership becomes the steward of alignment, ensuring that innovation moves fast but never outruns human values.

If intelligence is becoming the environment in which decisions are made, the natural question is whether any organizations are already operating this way. The answer is yes. Their experience offers a simple truth: The future is observable and already producing material advantages for those who adopt intelligence early.

What follows moves from principle to practice. First, we look at how intelligence is already operating at scale across critical industries. From there, the focus shifts to the operating models and leadership choices required to embed intelligence responsibly and competitively inside the enterprise.

Lessons from AI at Global Scale

Some industries are already operating in what others still imagine as the future. Industries defined by extreme complexity, like telecom, hyperscale cloud, and advanced life sciences show what it looks like when intelligence is no longer experimental but essential.

The first lesson is that breakthroughs emerge when AI integrates knowledge at scales humans cannot. In life sciences, platforms now analyze tens of petabytes of biological data and run millions of simulated experiments each week. As Dr. Azoulay, chief medical officer of a major cancer innovation institute, observed, "Breakthroughs emerge when AI integrates knowledge at scales humans cannot—moving intelligence from experimental to essential." Cycles that once took years compress into months. The same pattern appears in financial intelligence platforms, where markets are interpreted through continuous streams of filings, signals, and news.

The second lesson is that intelligent infrastructure multiplies resilience. Hyperscalers run self-correcting architectures that detect anomalies, predict failures, and rebalance workloads across millions of servers in milliseconds. Public benchmarks show significant uptime gains and dramatic cost reductions for certain incident classes.

The third lesson is that agentic AI transforms operations. Global telecom networks, serving billions and supporting 15% of global GDP, as estimated by the World Bank, now rely on autonomous agents to detect faults and act before customers notice. Across telecom and cloud environments, closed-loop systems adjust performance in real time, improving reliability and reducing energy use. Google's data-center intelligence autonomously optimizes cooling, workload placement, and energy consumption, showing how intelligence becomes infrastructure. Verizon's agent-to-agent AI detects anomalies, anticipates failures, and triggers corrective workflows without human intervention. These are not pilots; they are operating at global scale.

The fourth lesson is that intelligence becomes a new revenue layer. Across both developed and emerging markets, organizations are commercializing prediction itself: autonomous orchestration engines, industrial digital twins, AI-driven enterprise assurance, and real-time risk and decision models. Revenue shifts from products and access to the intelligence that improves them.

The fifth lesson is that AI delivers societal value at scale. During a panel I moderated at a global industry summit, an executive described how AI-powered 5G transformed a remote mining site in China into a near-autonomous operation. Instead of dozens of workers climbing a 4,000-meter mountain daily, a small team supervised operations from a city control center. AI optimized energy use, shifted power-intensive processes to off-peak hours, and reduced operational risk. Productivity rose, emissions fell, and safety improved, a vivid reminder that intelligent systems can elevate human wellbeing, not just enterprise performance.

Across these sectors, the pattern is unmistakable. Each cycle of sensing, reasoning, and acting makes the system more capable, more resilient, and more valuable than it was the day before. They show that once intelligence becomes embedded, value creation accelerates and operating models reorganize around a new logic of performance. The question is no longer whether this transformation will occur, but how quickly enterprises adapt to intelligence as a new layer of advantage.

Designing the Adaptive Operating Model

As intelligence becomes embedded in the organization, enterprises must redesign how they work long before rethinking the technology. An AI-native operating model is not an IT initiative but an architectural shift with a redesign of the operating model supported by four foundations.

The first is embedding intelligence directly into core workflows so that AI becomes a companion, advisor, and in some domains, an autonomous operative. When functions draw from the same cognitive foundation, the enterprise stops "running pilots" and starts operating on intelligence.

The second is creating continuous learning loops that allow the enterprise to sense, reason, and act in real time. These loops transform behavior from reactive to predictive. With each iteration, models sharpen, processes improve, and decision velocity increases. The organization begins to behave like a living system: self-correcting and capable of anticipating risks and opportunities before traditional cycles would detect them.

The third foundation is AI-native governance, where oversight shifts from periodic review to continuous assurance. Governance must be built into every step of the intelligence lifecycle, from data lineage and model validation to ethical guardrails and meaningful human judgment. Leaders become the stewards of alignment, ensuring intelligent systems behave predictably, protect trust, and support the organization's purpose and fiduciary duty.

The fourth foundation is investing in capability, not isolated tools. This means strengthening data, reusable models, human–machine collaboration, and shared intelligence platforms that compound over time. Such investments improve resilience, expand capacity, and enable adaptive operating models. Capital shifts from short term to building long-term architecture.

When these elements converge, the operating model becomes not just efficient but adaptive. Decisions accelerate, risks surface earlier, innovation cycles compress, and organization becomes more anticipatory.

A Practical Roadmap for Executives

As intelligence moves into the core of the business, leadership becomes the decisive factor in how well an organization adapts and competes. Small differences in data, architecture, and human–machine teaming can compound quickly. The leaders who win treat AI not as a deployment, but as a capability to steward and strengthen.

The first responsibility is readiness. No budget or ambition can compensate for an enterprise unprepared to absorb intelligence. Readiness isn't an IT checkpoint. It's organizational architecture: reliable data, standardized workflows, interoperable systems, strong governance, and teams able to work with intelligent systems. As MIT research shows, most generative AI pilots fail not because the models are weak, but because enterprises aren't structurally ready to operationalize them.

The second responsibility is to redefine how value is measured. AI creates meaningful P&L impact only when it changes how the business operates when it becomes part of core workflows and decision cycles. Yet many organizations rush to commercialize before establishing KPIs that reveal true enterprise value. Traditional ROI models, built for linear improvements, miss AI's deeper returns: capabilities that strengthen with use, platforms that scale across functions, and learning loops that continually refine decisions. Without this foundation, AI simply layers complexity to legacy architecture.

With it, AI becomes a business multiplier rather than a cost center, shifting investment from isolated projects to durable capabilities.

The third responsibility is to navigate the market with realism. The Wall Street Journal warns of circular investment patterns in the AI ecosystem: capital flowing from hyperscalers into startups and returning as cloud spend without durable end-user demand. Such loops can appear virtuous on the way up and vicious when momentum slows, echoing early internet bubble dynamics. For executives, the message is clear: Sustainable advantage comes not from chasing hype, but from building operational intelligence inside the enterprise.

The fourth responsibility is governance: embedded, continuous, and ethical. As intelligent systems influence everything from forecasts to societal outcomes, governance becomes the stabilizing force. Drift checks, transparency, human oversight, and ethical guardrails make AI trustworthy and scalable. Responsible AI isn't a barrier; it's the confidence layer that enables scale. Without trust, AI stays experimental; with trust, it becomes a strategic asset.

Leadership in the AI era is not about adopting technology but stewarding capability, ensuring that intelligence serves human values, organizational purpose, and long-term societal benefit. Thriving requires recognizing that the AI edge is the ability to build and apply intelligence that improves with use. Those who cultivate it early will define the competitive landscape for years to come.

Thriving as AI Reshapes Civilization

As AI weaves into every system of society, individuals face disruption unlike any before. Thriving begins with the mindset of seeing AI not as competition, but as amplification, when used with intention. Basic AI literacy becomes a new baseline of competence, and embracing this shift matters: Change is uncomfortable, but resisting it leaves people unprepared.

In an era of abundant machine intelligence, human-only abilities become even more valuable. Judgment, ethics, creativity, and narrative reasoning provide the direction algorithms cannot. These strengths shape both individual careers and the culture of intelligent enterprises.

Growth also requires continuous learning and choosing roles where AI enhances human contribution. Across society, AI is already expanding human potential: brain–computer interfaces giving voice to those who cannot speak, adaptive systems restoring mobility, and smart cities improving daily life. These breakthroughs show that AI does not diminish humanity, it extends it.

Ultimately, intelligent enterprises depend on individuals who develop not just AI literacy but AI wisdom, people who know how to collaborate with AI, question it, and guide it responsibly.

Building AI Wisdom

AI is a profound disruption, and the tension between its risks and its possibilities is exactly where its power lies. While Geoffrey Hinton, "AI grandfather," warns that AI could displace aspects of human labor or agency, this is not as a prophecy of catastrophe, but a reminder that intelligence without alignment can drift from human intent. He calls for responsibility, not fear.

At the same time, AI is not arriving as foreign intelligence that suddenly leaps beyond human reach. We evolve alongside it. Each day we build systems that help us see further, learn faster, and make better decisions, stepping on the shoulders of the intelligence we create. Whatever challenges emerge will not appear galaxies ahead of us, but only a step beyond where we already stand. That is precisely why the choices we make now matter so deeply.

The central truth remains: Technology alone does not shape the future, human judgment does. The future will be defined not by the intelligence we build, but by the wisdom with which we guide it. What individuals value shapes what enterprises build and what enterprises create shapes the trajectory of society. The future belongs not to those who simply build faster algorithms, but to those who architect a more resilient, capable, and human-centric civilization. The

Intelligent Enterprise is not the destination; it is the vehicle through which human potential is extended to global scale.

About the Author

Olga Martynov is a finance and sustainability executive with two decades of experience across high-growth startups, global industry consortia shaping AI and digital innovation, and a multibillion-dollar business within a Fortune 500 company. As a global CFO and advisor, she helps enterprises build the financial architecture, governance, and decision systems required to operate responsibly in the age of intelligent technology.

Born in Belarus, where freedom and opportunity were not inherited, she came to New York, where she built her career through resilience and reinvention, shaping her belief that adversity can be a catalyst for transformation.

A Harvard alum, speaker, and mentor, she is committed to human-centric leadership, responsible AI, and sustainable value creation, supporting leaders as they build organizations where people and intelligent systems advance together.

Email: olga.martinov@gmail.com

LinkedIn: https://www.linkedin.com/in/olgamartynov/

BEYOND MEDICINE: ARTIFICIAL INTELLIGENCE, ROBOTICS, AND THE HUMAN BRAIN IN DEEP SPACE EXPLORATION

By Agnieszka Mietz-Blijleven, PhD
Global Head of Architecture Engineering, AI
Osielsko, Poland

Nothing in life is to be feared, it is only to be understood. Now is the time to understand more, so that we may fear less.

—Maria Skłodowska-Curie

After completing my PhD, "Artificial Intelligence with Charge Transfer in Deep Brain Stimulation: New Approaches Based on AI/ ML, LLM, Generative AI, and Cybersecurity in the Diagnosis and Treatment of Brain Disease," which explored the intersection of advanced computational models and quantum physics, particularly charge transfer mechanisms in neural modulation, early detection of ischemic strokes, and cortical mapping in migraine with aura, I found myself asking a fundamental question: Is this the moment to truly explore the operational potential of the human brain within the vastness of the universe?

We are standing on the threshold of civilization's next great disruption. Artificial intelligence is no longer merely a computational tool. It is becoming the architect of post-biological evolution. When combined with robotics, nanotechnology, and exascale computing, AI offers humanity an unprecedented opportunity to transcend the limitations of carbon-based biology and reach the deepest, most unexplored regions of space.

Artificial Intelligence and Robotics: The Vanguard of Space Autonomy

In the hostile environments of deep space, where cosmic radiation, microgravity, and isolation degrade human physiology, AI-driven robotics emerge as essential allies. Powered by generative AI and large language models (LLMs), these systems act as autonomous agents capable of adapting to unknown planetary conditions. Through multimodal sensor fusion, Bayesian inference, and reinforcement learning, they interpret complex stimuli and make decisions in real time. Their bodies, composed of programmable nanomaterials and shape-memory alloys, enable morphogenesis and self-repair, ensuring resilience in unpredictable terrain.

Yet their role extends beyond mechanical function. Robots can interface directly with the human brain via non-invasive brain-computer interfaces (BCIs), decoding motor intentions and emotional states using EEG, fNIRS, or intracortical signals. In doing so, they become cognitive surrogates—extensions of human thought and emotion—co-evolving alongside their human counterparts. This is not automation; it is symbiosis.

Charge Transfer and Brain-Machine Interfaces

The quantum realm offers profound insights into cognition. My research into charge transfer dynamics revealed that the stochastic behavior of electrons in ionized plasma mirrors the probabilistic firing of neurons. This parallel opens new possibilities for brain-machine interfaces that operate in harmony with quantum phenomena.

Imagine closed-loop deep brain stimulation (DBS) systems that use AI to modulate neural oscillations in real time, enhancing cognitive performance and emotional regulation during long-duration missions. Neuroadaptive suits embedded with electroactive polymers and biosensors could respond to stress signals from the cortex and autonomic nervous system, adjusting support dynamically. Transcranial magnetic stimulation (TMS) guided by AI could deliver targeted nootropics, boosting memory, executive function, and psychological resilience.

Even spacecraft interiors could be designed to resonate with neural rhythms, adjusting lighting, acoustics, and task pacing based on real-time brainwave analysis. In this vision, the environment itself becomes neuroresponsive, attuned to the mental states of its inhabitants.

Onboard Supercomputers: The Silent Revolution

The integration of onboard supercomputers marks a turning point in space autonomy. Despite challenges such as heat dissipation and radiation exposure, advances in cryogenic cooling, photonic processors, and neuromorphic chips are making it possible to embed intelligence directly into spacecraft infrastructure.

These systems can simulate alien biospheres using multi-agent ecological models and evolutionary algorithms, preparing astronauts for unknown ecosystems. They can predict quantum anomalies in navigation using entanglement sensors and topological error correction, ensuring precision in interstellar travel. They can also host LLMs that serve as multilingual translators, mission strategists, and affective computing agents, detecting crew stress and initiating psychological interventions.

In this paradigm, the spacecraft is no longer a passive vessel. It becomes a sentient habitat: responsive, adaptive, and deeply integrated with the cognitive and emotional needs of its crew.

Biological Barriers and the Future of Space Suits

Human biology remains the most fragile component of space exploration. Galactic cosmic radiation (GCR), hypoxia, and psychological isolation pose existential threats to long-term survival. To overcome these barriers, we must reimagine the space suit, not as armor, but as an extension of the nervous system.

Cybernetic exosuits constructed from nanostructured metamaterials could provide dynamic radiation shielding, thermal regulation, and pressure stabilization. Hybrid EMG/EEG arrays would allow the suit to anticipate movement and reduce muscular fatigue. Embedded AI diagnostics could monitor biomarkers such as cytokines, cortisol, and neurotransmitters, delivering microdoses of medication or stimulation via transdermal or optogenetic systems.

Visual functions would be enhanced through intelligent ocular lenses that adjust to pressure, focus, and light exposure. These suits could forecast health impairments using neurophysiological imaging and feedback loops, intervening before symptoms manifest. In essence, the suit becomes a second skin: intelligent, responsive, and biologically integrated.

Into the Abyss: Black Holes and Undiscovered Worlds

As we contemplate missions beyond the heliopause toward rogue planets, interstellar voids, and even black holes, we must rethink the architecture of consciousness itself. Near a black hole, spacetime curvature, gravitational time dilation, and tidal forces render traditional biology obsolete. Survival may require substrate-independent minds uploaded into quantum-hardened robotic vessels.

Navigation could be guided by gravitational lensing, using the black hole's curvature to slingshot probes across vast distances. AI systems would analyze hyperspectral data to detect biosignatures in the atmospheres of unmapped ocean worlds, carbon-rich super-Earths, and tidally locked exoplanets orbiting red dwarfs.

These missions are not merely scientific. They are philosophical. They challenge our understanding of identity, embodiment, and intelligence in a universe that may harbor life forms beyond our imagination.

Final Reflection

As we push toward Mars, Venus, the Moon, and beyond into the gravitational shadows of black holes and the spectral silence of undiscovered planets, the question is no longer "Can we go?" but "How will we evolve to thrive?"

The convergence of AI, robotics, quantum physics, and neuroscience is not just a technological revolution. It is a philosophical metamorphosis. It challenges us to redefine cognition, embodiment, and identity in a cosmos that is vast, dynamic, and increasingly within reach.

We are not merely building spacecraft. We are building cognitive ecosystems where brain, machine, and universe coalesce into a new form of sentient exploration. This is not the end of human evolution. It is its expansion. A moment where consciousness becomes interstellar, where biology meets quantum substrates, and where the human story is rewritten across the stars.

About the Author

Dr. Agnieszka Mietz-Blijleven is an international manager, architect, MCT, DevOps trainer, and Microsoft MVP in Azure AI Services. She has for many years served as a manager within global structures of managers and architects, while simultaneously working hands on as a cloud solution architect specializing in AI/ML and cybersecurity. She holds degrees from eight universities across Poland, Germany, the UK, the USA, and Italy, with expertise spanning computer science, economics, commodity engineering, and medical sciences.

Her PhD research focused on artificial intelligence with charge transfer in deep brain stimulation, exploring intersections of AI/ML, generative AI, and cybersecurity in diagnosing and treating brain diseases. As a trainer and consultant, she leads projects in AI, DevOps, and cybersecurity, specializing in highly confidential and regulated environments. She is passionate about transforming complex technical processes into clear, reproducible educational materials and advancing the integration of AI and neuroscience in space exploration.

Email: Agnieszka.mietz@gmail.com

LinkedIn: https://pl.linkedin.com/in/agnieszkamietz

CHAPTER 17

RECLAIMING THE LEADER: HOW AI RESTORES STRATEGIC COMMAND

By Tevan Millette
Founder, Momentum Mastery™ Consultant, Coach
Destin, Florida

Beware the barrenness of a busy life.

—Socrates

There is a particular kind of quiet that surrounds a person who has been strong for too long. Not the quiet of peace, but the quiet of numbness, the silence you slip into when you are holding everything together but no longer remember why.

I remember a night like that. It was not dramatic. Not catastrophic. Just still. The house was asleep. My phone, for once, wasn't ringing. The task list for tomorrow was already longer than the day I had just survived. I sat there, jaw tense, pulse steady, mind full, and felt the faint realization that I had slowly become someone I never intended to be. Not weak. Not broken. Just buried. Buried in execution. Buried in solving. Buried in serving. Buried in being indispensable.

Somewhere along the way, I had stopped leading and started holding. Somewhere along the way, I had become the machine I built. And that is the quiet that breaks a person. Not the noise of overwhelm, but the subtle erosion of identity. The truth that surfaced in that stillness was simple: I wasn't exhausted because I was doing too much. I was exhausted because I was doing what did not require me. My identity had been traded for efficiency. And I had done it to myself.

This chapter is about that exchange and how we take ourselves back.

The Real Reason Leaders Burn Out

Most entrepreneurs, executives, creators, and founders are not overwhelmed because they lack skill, discipline, or clarity. They are overwhelmed because they are drowning in responsibilities they were never meant to own in the first place. We build the thing. We protect the thing. We maintain the thing. We feed the thing. And then one day, we become the thing.

We become the operator, the systems layer, the firefighter, the bottleneck, the emotional container, the strategist, and the executor, all at once. We start our journey as visionaries, creators, builders of outcomes. But somewhere along the path, we transition into something much smaller: We become the engine. We move from:

The Visionary	The Operator
From designing outcomes	To reacting to needs

The Visionary	The Operator
From leading	To maintaining
From creating	To surviving

Burnout isn't the result of effort. Burnout is the result of identity displacement. When you forget who you are in the system, you begin working against yourself from inside it. And that is where AI enters this conversation. Not as a technological miracle, but as an existential invitation.

For years, I believed exhaustion was the proof of my commitment. Long days and longer nights were badges of honor. I wore them proudly until I realized those badges were made of lead. They didn't make me stronger; they were slowly pulling me under. Leadership had become a performance of endurance rather than a practice of intention. Somewhere along the way, I confused momentum for meaning. I was busy, constantly in motion, but directionless like running on a treadmill convinced I was conquering mountains. Most leaders don't notice the shift when it happens. It's quiet.

The emails keep coming, the team keeps asking, the meetings keep multiplying. You tell yourself it's temporary, just one more push until the next milestone. But milestones become marathons, and marathons become lifestyles. Before long, you're standing at the summit of your own making, exhausted and strangely hollow.

The great irony is that the very traits that make you successful—resilience, ownership, and self-reliance—are the same ones that trap you in the cycle. You keep saying yes because you can. You keep carrying because you always have. You keep pushing because that's what leaders do. But leadership isn't about how much weight you can bear. It's about what you choose to carry and why.

When you stay buried in execution long enough, you forget that your greatest value isn't in the doing; it's in the designing. Your team doesn't need your exhaustion; they need your clarity. Your business doesn't require your proximity; it requires your perspective. That realization is sobering because it forces you to confront a truth most leaders avoid: We often build systems that depend on our dysfunction. We build organizations that need our constant intervention, because the chaos makes us feel important.

We create dependency loops where our identity is validated by being needed. And when those systems finally grow beyond our reach, we feel lost and not liberated. It's not that we're afraid of being replaced. We're afraid of being free. Because freedom means facing the silence that work used to drown out. That's why burnout isn't just fatigue, but rather its identity crisis disguised as productivity. It's the moment when you look at everything you've built and realize you're nowhere to be found inside it. That's where I found myself that night: the hum of success around me but no sense of self within it.

That's why I call AI an existential invitation. Because beneath all the code and capability, it poses a profoundly human question: "What if you could remove everything that doesn't require you, would you still know who you are?"

AI forces us to revisit the architecture of our own value. It asks us to decide what truly needs our humanity and what can be released. This isn't about learning prompts or chasing the next tool. This is about reclaiming design authority over your life and work. Imagine a leader who's constantly firefighting. Their day begins with urgency and ends with depletion. They manage people, but mostly manage chaos. They call it leadership, but it's survival.

Now imagine that same leader stepping into a new rhythm where AI handles their daily reports, monitors project progress, summarizes meetings, drafts initial communications, filters decisions, and prepares analysis before they even arrive. Suddenly, they have *space.*

What do most leaders do with that space the first time they taste it? They fill it. They find new fires. They overcompensate. Because space feels foreign. That's where your evolution begins. Not in learning how to use the tool, but in learning how to tolerate the quiet it creates.

When the noise clears, you're left with questions that can't be delegated: Who am I as a leader when I'm no longer defined by busyness? What does value look like when effort is no longer the currency? How do I lead people through technology without losing our humanity? These questions mark the threshold between the operator and the architect.

Every era of disruption offers a choice: resist, react, or redesign. AI is simply the latest and perhaps the most significant invitation to redesign. Leaders who resist will drown in the speed of change. Leaders who react will automate chaos. But leaders who redesign will thrive.

Redesign begins by remembering the difference between control and command. Control is reactive. It's micromanagement, constant correction, perpetual urgency. Command is calm clarity. It's vision, coordination, orchestration. AI gives leaders the capacity to return to command. It doesn't diminish leadership; it refines it. It removes the noise so the signal can finally be heard.

When leaders embrace that shift, something almost spiritual happens: Meetings become conversations again, not firefights. Creativity returns because the mind finally has room to wander. People begin to follow vision rather than process. Culture strengthens because burnout weakens. And in that regained clarity, leaders realize that technology was never the enemy. It was the mirror. A mirror reflecting how much of their life had been spent doing what could have been designed instead.

That's why I believe AI isn't the end of leadership. It's the rebirth of it. The leaders of the next era won't be defined by how well they use technology. They'll be defined by how well they partner with it. They'll know that power doesn't come from output but from orchestration: the art of aligning people, purpose, and platforms around a unified direction. That's the new definition of scale: not more effort, but amplified essence. AI is the mechanism; identity is the multiplier. When the two align, the modern leader stops chasing efficiency and starts cultivating capacity. And that is where the future of leadership truly begins.

The Misunderstanding About AI

Most conversations about artificial intelligence fixate on the wrong question: Will AI replace human beings? The truth is more nuanced, more human, and more personal: AI will replace the parts of ourselves we were never meant to be. The busywork. The constant context switching. The repetitive cognitive cycles. The administrative drag. The information processing and triage. AI is not here to take our place. It is here to take our burden. The real disruption is not artificial intelligence replacing human intelligence.

The real disruption is this: AI separates those who know how to use leverage from those who only know how to use effort. For the first time in modern history, the advantage is shifting away from brute-force output.

The world is beginning to reward architects, not engines. Creators of systems. Designers of flow. Leaders of direction. The era of the self-sacrificing operator is ending. The era of the strategic leader is returning.

AI as a Capacity Multiplier

When used with intention, AI becomes something far more powerful than automation.

It becomes a multiplier of identity. Here is what AI does when used properly and humanely:

- Reclaims bandwidth, freeing your mind to think rather than react

- Reduces cognitive load, decreasing decision fatigue and emotional drain

- Amplifies creativity, expanding the space for experimentation, innovation, and vision

- Restores presence, allowing leaders to re-engage with the human parts of life and work

- Enables scale without self-sacrifice, breaking the false belief that growth must cost your wellbeing

AI does not make you more productive. AI gives you your mind back. And once your mind returns, so does your leadership.

The Identity Shift: From Engine to Architect

The moment a leader realizes they are not meant to be the machine is the moment transformation begins. The core shift is this: Stop being the person who does the work.

Become the person who designs how the work gets done. This is not laziness. This is leadership.

This is how nations are led, how missions are executed, how meaningful movements are built. It is how you reclaim:

- Strategic thought
- Emotional clarity
- Creative energy
- Personal presence
- A grounded sense of self

AI is not a tool to help you work harder. AI is the tool that allows you to return to your natural altitude.

The Capacity Reclamation Framework™

This simple, human-centered model sits at the heart of my work with leaders and organizations. It's designed to help people reclaim the bandwidth, creativity, and clarity they've lost to over-execution.

- Remove: Identify tasks, decisions, and responsibilities that do not require your judgment or identity.
- Systemize: Standardize anything that repeats; patterns are leverage.

- Delegate to AI: Not to replace thinking, but processing
- Reserve the Human: Protect your finite energy for what only you can do: vision, judgment, communication, and relationship.

These four principles form the backbone of what I call Capacity Reclamation™, a practice leaders can begin applying today. A deeper exploration of this model forms part of my forthcoming work, where I expand on how AI can help humans build sustainable capacity, clarity, and leadership in an increasingly complex world.

When the Leader Returns, Everything Changes

As AI removes the noise, something unexpected happens: You begin to think again. Ideas resurface. Curiosity reopens. Conviction returns. Your pace slows, but your impact increases. Your presence deepens. You stop reacting and start designing.

You stop surviving and start leading. And the irony is profound: The more you let AI carry the weight, the more human you become.

This Is Not About Technology

This chapter is not about learning prompts or tools or software workflows. This is about identity. Who are you when the noise stops? Who are you when you are no longer drowning? Who are you when your time is no longer mortgaged to urgency? Who are you when you are no longer defined by effort?

There is a version of you that remembers:

- Why you started
- Who you are meant to lead
- What you are here to build
- And the impact only you can create

AI did not give you that. AI simply cleared the path back to it.

The Leader Returns, Not as Who They Were, but Who They Always Were

The night I sat in that quiet, I didn't make a declaration or write a plan. I just remembered. I remembered that leadership is not measured by how much you carry.

It is measured by the clarity of what you choose to hold. I remembered that identity is not something we earn. It is something we return to. I remembered that strength is not in endurance. Strength is in alignment. AI did not change who I am. It gave me the space to become him again.

About the Author

Tevan Millette is a Special Forces Green Beret, transformational leadership coach, AI consultant, Maxwell Leadership Certified Member, and founder of Momentum Mastery™, a keynote, coaching, and training platform that helps people and organizations unlock their full potential. Known for empowering both high-performing leaders and everyday achievers to reclaim identity, clarity, and sustainable drive, his work bridges high-performance leadership, emotional resilience, and systems-based growth. Through his certified coaching practice, courses, consulting, and speaking programs, Tevan guides entrepreneurs, executives, and individuals seeking meaningful change out of reactive, burnout-driven patterns and into intentional, purpose-driven leadership. While his AI consulting equips businesses to scale intelligently, his broader mission is to help people build lasting momentum and create lives and organizations that are not only successful but deeply aligned, human, and impactful.

Email: CustomerService@TevanMillette.com

Website: www.TevanMillette.com

LinkedIn: https://www.linkedin.com/in/tevan-millette

CHAPTER 18

GUERILLA AI—A PLAYBOOK FOR SMBS

By Gerald Pallor
Owner, Insight Media; Publisher, Byte Sized Tech Tips
Newburgh, New York

Beauty is truth, truth is beauty, that is all Ye know on earth, and all Ye need to know.

—John Keats

This chapter focuses on marketing, not because marketing is glamorous, but because it is where artificial intelligence delivers the fastest, clearest advantage for small and mid-sized businesses. Marketing has always favored the organizations with the biggest budgets, the widest reach, and the most polished teams. AI changes that balance. It gives SMBs the ability to act with the precision, speed, and creative force once reserved for large agencies.

I call this guerilla AI: the strategic use of AI to compete above your weight class. It does not replace people. It removes friction, sharpens judgment, and gives small and mid-sized firms leverage that once required large budgets. In marketing, where effectiveness is measured only by audience response, that leverage matters.

This chapter outlines some strategic and tactical advantages AI brings to SMBs. It does not attempt to cover every aspect of marketing. Instead, it focuses on moves that give these teams the greatest lift: clarifying strategy, defining identity, producing creative, placing media intelligently, and using data to learn what works. These are the parts of marketing where AI creates the kind of asymmetry an SMB can turn into competitive strength.

AI Makes Precision Affordable

Traditional marketing favors large companies. They have the budgets, the agency talent, the production resources, and the media reach to shape perception and hide mistakes. AI flips that equation. It gives SMBs access to strategic clarity, creative production, and measurement at a fraction of the historical cost. It allows them to compete in a crowded landscape with precision instead of spending.

The truth of a campaign appears only in its results. When people move, the message holds and the business grows. When they ignore it, the claim fails. A campaign that generates no response remains unfinished because the audience never completes it. The governing truth of all marketing is simple: It succeeds when audiences respond. Beauty and truth converge in results.

Flipped Interaction: A Strategic Superpower for SMBs

Guerilla tactics depend on breaking habitual patterns. Most people ask large language models (LLM) direct questions and accept whatever answer they receive. Flipped interaction inverts that approach. Instead of asking for an answer, you state a goal and ask the LLM what questions must be answered to reach it: not "What happens if I double my ad spending?" but "What questions do I need to resolve to make my advertising more effective?"

A young businessman from a wood-molding company once asked me what AI could do for him. I asked a single question: Are you involved in international trade? His answer—no—eliminated export risk, tariff exposure, and global pricing pressure. The entire strategy shifted to local craftsmanship, design quality, reliability, and speed. One clarifying question rewrote the plan.

Had he gone straight to AI for a marketing plan, he would have received something generic. Flipped interaction prevents that. You describe what you do, state your goals, and let the model surface the unknowns. It forces clarity. It exposes blind spots. That is where strategy begins. Once the strategy is defined, the work moves to brand identity, creative execution, media placement, and data analytics.

Identity: Fast, Clear, and High-Quality

Branding defines a business's identity and the message its marketing must express. AI does not create this; it helps owners articulate it. Effective branding begins with positioning: a one-line claim—the unique selling proposition—that establishes the competitive stance and guides every creative decision. From that single line come the promises and messaging pillars that every piece of communication must reinforce.

AI drafts, tests, refines, and tightens these statements quickly. Once the foundation is set, the creative brief follows: logos, fonts, and color systems. The difficulty is visualizing how these elements carry across all media. AI tools assist by generating high-quality logos, palettes, font sets, and asset kits, along with mockups for business cards, web banners, and social posts. Be sure to review everything with a professional before deploying.

Creative Production Without a Studio

After branding elements are established, content production begins. Campaign creative has four functional areas: copywriting, graphic design, motion content, and audio. Treat each area as a separate discipline even when one person handles them all.

Copywriting

LLMs accelerate drafts, rewrites, tone adjustments, and strategic alignment, but the human still makes every decision. Use the model as a collaborator: Draft, critique, and refine until the copy matches the strategy.

Graphic Design

AI generates images, layouts, and illustrations from prompts, and it evaluates composition, hierarchy, spacing, color, and legibility. This lets SMB teams raise design quality without hiring specialists.

Video and Animation

AI video tools create explainers, promos, and social clips from text or images, giving small firms a workable rough edit without a production crew. Traditional fine editing is often required for timing and continuity, but AI reduces the cost of reaching the first stage.

Audio

Voice tools provide consistent narration and can clone a brand voice; music tools generate background tracks or simple themes. They help maintain cohesive sound across campaigns, though complex narration or long-term brand jingles still benefit from human talent.

Smart Media Placement, Less Guesswork

The creative isn't worth a dime if it isn't seen, and the more it is seen in high-quality outlets, the better. Media strategy determines where the company's creative will show up, how often, with what budget, in what sequence, and for which market segments.

Media channels fall into three groups: owned (website/email), earned (SEO/organic social), and paid (ads, programmatic). Search engine optimization (SEO) affects discoverability, determining whether your content appears when people search your category. It must be planned alongside all other placements because the content you publish—articles, guides, posts, pages—must support the brand message and the campaign goals.

There is still a world beyond digital. Printed postcards, flyers, or ads in local papers, local cable, direct mail, billboards, and other out-of-home channels are traditional channels that are still effective.

Any store owner will tell you that the most important marketing element is the sign outside their building. Without it, nobody will find you. In-store signage is critical, even in a small store.

AI tools act as the planning and optimization engine inside media strategy. It analyzes audience behavior, compares channel options, allocates spending across placements, recommends cadence, forecasts results, and adjusts activity based on performance.

AI-Assisted Analysis and Disciplined Data Hygiene

Professional marketers use AI to interpret evidence, not just generate text. The advantage comes from turning raw data into actionable insight.

RFM Analysis (Recency, Frequency, Monetary Value)

RFM analysis identifies your most valuable customers based on when they last bought, how often they buy, and how much they spend. It produces segments for targeted promotions and forecasts buying cycles and warns when customers begin drifting toward churn.

Market Basket Analysis

Market basket analysis reveals which products tend to sell together. It informs shelf placement, cross-promotions, bundling, and ordering decisions.

Segmentation

Segmentation groups buyers by shared behavior or characteristics. It improves media planning, message alignment, and targeting.

Sentiment Analysis

Sentiment analysis reads how customers actually feel about your brand. It guides messaging, highlights strengths, and flags emerging problems.

Regression Analysis

The core tool of analytical marketing is regression analysis. It describes relationships, predicts outcomes, and prescribes actions by quantifying which variables drive results. It supports causal inference, anomaly detection, and risk estimation—essential management capabilities.

Data Discipline

Clean, uniform data is essential for all analytical techniques. Garbage in, garbage out. SMBs don't need data scientists; they need consistent input of relevant, properly structured information.

The first step is to stop improvising. Freeze the inputs before cleaning anything. Define the categories you actually use and eliminate the rest. Force every entry into fixed formats. Clean the existing data once: Remove duplicates, correct dates, standardize names, fix numerical fields, and drop corrupted entries. Then lock the structure and keep it stable.

It takes time, but it pays off. AI can guide the cleanup; discipline maintains it. The business prospers from the clarity that follows.

The Role of Humans: Judgment, Taste, and Reality Testing

Every marketing effort still rests on three fundamentals: Understand your customers, measure their behavior, and speak to them persuasively. AI can support each step, but it cannot replace human reasoning. Never treat an LLM as the final authority. It calculates from patterns; it does not think and it does not understand.

Approach AI creative tools the same way you approach any creative project: Execute, critique, revise, and refine until the work is ready. The distance between playing with prompts and running a campaign is the distance between guessing and knowing. AI makes it easy to start; expertise makes it work.

There is fear that AI will take jobs from people. In practice, AI is more likely to increase employment in a growing SMB. Processes change, new demands appear, and the work expands. The roles may shift, but the need for people remains—and often grows.

In the marketplace, truth and beauty are measured by response. The audience decides whether the work succeeds or fails. AI can assist at every step, but it cannot do the job for you.

Never forget: The human brain is a shape-shifting, self-repairing, hyper-parallel system running on 20 watts. AI models are large calculators that require a data center to rest between training cycles. Advantage humans.

Guerilla AI as Competitive Equalizer

SMBs are more agile than the enterprise scale businesses that dominate the market, and they do not need to dominate to be successful. AI strengthens the three competitive advantages SMBs have: speed, clarity, and adaptability.

The goal is to use AI to make better decisions, produce stronger creative, and find truth through audience response.

Always remember: AI accelerates, but humans finish the work.

About the Author

Gerald Pallor began his career as an independent film and video producer, creating news segments and documentaries. He later studied direct marketing and used those methods to define audiences, target them precisely, and deliver persuasive messages for businesses, nonprofits, and institutions. He added data analytics to measure campaign performance and project outcomes with greater accuracy.

He turned his attention to artificial intelligence in 2024 and launched the Byte Sized Tech Tips newsletter on Substack. That work established him as a fractional CMO and technical consultant to small and mid-sized businesses.

His spare time is spent in the gym, having lunch with his adult son, and sitting on his porch overlooking the Hudson River, marveling at how fortunate he has been.

© 2025 Gerald Pallor. All rights reserved.

Website: www.geraldpallor.substack.com

LinkedIn: https://www.linkedin.com/in/geraldpallor/

REIMAGINING LEARNING: GENAI AND THE FUTURE OF EDUCATION AT SCALE

By Russell Ramesh
EdTech Leader & GenAI-in-Learning Pioneer
Baltimore, Maryland

Innovation in education is not about building smarter machines, but about building tools that make humans wiser.

—Steve Jobs

The Great Inflection Point

Education today stands at an inflection point that rivals the invention of the printing press or the advent of the internet. The rise of generative artificial intelligence (GenAI) has altered the grammar of learning itself. It has blurred the once-rigid boundaries between teacher and learner, content and context, and human cognition and machine capability.

As one who has spent decades leading EdTech organizations and advising the US Department of Education's Office of Educational Technology (OET), I have witnessed the sector's evolution through several waves: from digitization to personalization, from online access to adaptive intelligence. Yet, none has held the transformative potential of GenAI. Unlike prior tools that merely digitized existing pedagogies, GenAI redefines the act of learning by co-creating knowledge with the learner.

We now live in an era where a student's first tutor might be an algorithm, yet their deepest insights may still arise from human mentorship. The challenge before us is not to choose between man and machine, but to orchestrate both in harmony, an academic symphony tuned to the needs of the individual learner and the demands of the age.

When I first read that Aristotle had personally tutored Alexander the Great for 14 years, from the age of two to 16, I remember feeling a twinge of envy, an almost childlike jealousy. I thought to myself how extraordinary it must have been to learn directly from a mind that shaped the foundations of Western philosophy. Through the miracle of the printed page, I can at least access what Aristotle inscribed two millennia ago, unfiltered and unmediated, perhaps supplemented by a good philosophy professor who adds interpretation. Yet, I cannot ask Aristotle a question. I can ask, but I will not get an answer.

My hope, and the quiet mission of this decade, is that we may soon fashion a mainstream educational tool that changes this dynamic: an interactive Aristotle, brought to life through generative AI. In my years at the OET and in the industry, I have seen data transform education; data-driven learning was the last great revolution, democratizing access and measuring outcomes with unprecedented precision. But now, we are entering a new era, one I often call the era of free intellectual energy.

This energy is powered by data but driven by GenAI. It is still crude but growing more refined with every passing month. In our lifetime, it will become exquisitely finessed. My aspiration is that when the next Aristotle, or a 21st-century Socrates, emerges, we will be able to capture not only their teachings but their worldview, reasoning patterns, and moral temperament in digital form. Someday, a student may not only read Aristotle's words but engage in a true dialogue with them.

As long as AI amplifies the learner's inherent ability, it remains a force for good. It can lift students out of mechanical tasks and free them to operate at the conceptual and creative levels where human intelligence shines. It can also deepen the relationship between learner and instructor. This shift will not take generations. It will unfold within the next decade, perhaps even sooner, permeating classrooms and communities and giving every learner an opportunity that once required an Aristotle. The truth at the center of this inflection is simple: Technology matters only when it strengthens what makes us human, and GenAI is the first tool capable of revealing rather than replacing our intellectual essence. It is becoming the modern bicycle for the mind, offering every learner a lift that was once reserved for the fortunate few.

From Content Delivery to Cognitive Partnership

For decades, technology in education was a tool of convenience. It made content faster to distribute, easier to measure, and more engaging to consume. Yet it did not change the essence of learning. It improved the textbook without reimagining the classroom.

Generative AI has changed that equation. It transforms content into cognition, it converts learning from a one-way transfer into an active conversation. For the first time, we have a technology that can mirror curiosity, respond to emotion, and collaborate with thought itself.

I often recall an experiment from my school years that shaped how I think about the relationship between humans and tools. It was called "Locomotion Exploriment." In those days, computer labs were almost sacred. We entered without shoes, without food or drink, and with an unspoken sense of awe. The simulation asked us to measure the efficiency of movement across species, expressed in kilocalories per gram per kilometer. Humans walking barefoot performed sub-optimally. The salmon glided through water with astonishing grace, and the Andean condor traveled the farthest on the least energy.

Then came the pivotal moment. When you placed the human on a unicycle, performance improved modestly. But once you put the human on a bicycle, everything changed. The unaided walker expended about 0.75 calories per gram per kilometer, while the cyclist required only 0.15. The human, assisted by a simple machine, became the most efficient mover of all.

That lesson has stayed with me for decades. Humans are not the fastest or strongest beings on the planet, yet we possess the unique ability to create tools that amplify our natural capacity. The bicycle did not change what it meant to be human; it revealed what was possible when innovation met intent.

Education technology works on the same principle. The analog textbook represents the unaided human, noble in purpose but limited in reach. Digital learning platforms placed the learner on a unicycle, allowing for incremental improvement. When we added adaptive feedback, interactive assessment, and data-informed instruction, we put the learner on a bicycle. And now, with AI in EdTech, we are approaching the equivalent of a fast car that still keeps the human in control while extending the range, speed, and precision of the journey.

AI does not simply automate content delivery. It builds partnership. It learns alongside the learner, senses context, and refines understanding. It allows students to move beyond mechanical recall into realms of inquiry and creativity. It frees educators from repetitive tasks so they can devote their time to guiding, inspiring, and cultivating imagination. At its best, AI serves as a cognitive accelerator. It channels the same principle that powered the cyclist's leap in efficiency: a fusion of human purpose and engineered precision. It helps learners travel farther with the same effort and allows teachers to focus on what only humans can do: nurture empathy, curiosity, and wisdom.

EdTech is still in its early ascent, but the trajectory is clear. The goal is not to build machines that learn faster than humans, but to help humans learn more deeply because of machines. When technology becomes a true cognitive partner, education shifts from what is taught to what is discovered.

The idea that unifies this chapter is simple: AI becomes transformative only when it elevates human understanding rather than automating it. The bicycle once showed us how a simple tool can expand human capacity. GenAI now does the same for cognition.

The Compass and the Code

The Kiwi with Bionic Wings: A Parable for Purposeful Innovation

My father often told me a story that, over time, became a compass for how I approach technology in education, "The Kiwi with Bionic Wings." In the story, a well-meaning group of scientists sought to help a kiwi, a flightless bird native to New Zealand, experience the joy of flight. They designed prosthetic wings, conducted surgeries, and trained it in simulators. For a while, the kiwi struggled bravely. But soon it grew weary and disoriented, for its very nature was never meant for flight. It was born to run, not soar. The moral was clear: When innovation ignores essence, it breeds suffering, not progress.

This parable resonates profoundly in the GenAI age. Many institutions rush to "bolt AI wings" onto existing systems, adding chatbots to outdated curricula, replacing teachers with unverified models, or equating automation with advancement. They risk creating digital kiwis: systems that appear modern yet lose their native rhythm of learning.

True innovation in education must respect the organic nature of human development. GenAI should not force flight upon learners but extend their reach along the ground they are meant to cover, i.e., strengthening curiosity, autonomy, and confidence in their stride. Purposeful innovation is not about doing everything differently; it is about doing the right things differently.

The Arithmetic of Gratitude: A Philosophy for AI-Age Leadership

My father's second lesson was what he called "The Arithmetic of Gratitude." He believed that life multiplies blessings for those who count them, and divides burdens for those who share them. This arithmetic applies elegantly to education.

In the age of GenAI, gratitude translates to mindful design, to acknowledging the accumulated wisdom of educators, psychologists, and students who came before us. Gratitude reminds us that technology stands on the shoulders of humanity. Every model we train is, in essence, a reflection of collective human thought. For leaders, the arithmetic of gratitude also means sharing the dividends of innovation. When algorithms personalize learning for one student, they must also enrich the ecosystem for many. Education, at its core, is not a zero-sum equation. A rising tide lifts all boats.

The lessons from the kiwi and "The Arithmetic of Gratitude" converge on a single truth: Progress without purpose is motion without meaning. Technology should never exist for its own applause. Its purpose is to amplify human potential. The kiwi reminds us to innovate in harmony with nature and need. Gratitude reminds us that every advance stands on shared wisdom and collective effort. Together they form the moral architecture of the GenAI age, where success is measured not by how intelligent our systems become but by how humane our learning can be. The unifying theme is clear: Innovation must honor human nature, because the purpose of AI is not to impose intelligence but to awaken it. Even the finest bicycle cannot make a bird fly. Tools must serve natural learning rather than distort it.

GenAI and the Evolution of Pedagogical Design

Building an EdTech product requires holding thousands of interconnected concepts in your mind at once. You must see how cognitive science connects to user experience, how assessment logic interacts with data pipelines, and how motivation design ties into pedagogy. It is an intricate puzzle of form and function. Every hour brings a new variable, a fresh constraint, or an unexpected opportunity to fit the pieces together in a slightly better way.

That process, the integrity of the product development lifecycle, is what separates a passing trend from a timeless platform. It is the rhythm that drives innovation in education. My passion has always been to build an enduring company where people are as motivated to create great educational products as they are to sustain profitability. Profit matters because it fuels purpose, but purpose is what gives profit meaning. I spent circa two decades managing profit centers, and I learned that money is the oxygen of innovation. It keeps the system alive, but it is not the reason for being. What sustains a great EdTech company is not the chase for valuation but the devotion to craft, the belief that thoughtful design, aligned with pedagogy and powered by AI, can help learners discover the best version of themselves.

In the end, GenAI will not replace the designer, the engineer, or the educator. It will refine their craft. It will help us navigate complexity and uncover insights that were once invisible, yet the heart of every product will remain human. Great ideas will continue to matter, but it is the process, the craftsmanship, and the will to create something truly useful that turn ideas into lasting impact and fuel the next generation of upskilling and reskilling.

The central idea of this chapter is constant: AI reaches its highest value when it amplifies the educator's intent and the learner's potential. Designing GenAI-powered instruction is the act of improving the bicycle itself so that every turn of the pedal multiplies the learner's momentum.

The Architecture of Human-Machine Collaboration

Very early in my journey as an EdTech executive, I noticed something that puzzled me at the time. I could not explain it then, but I have reflected on it for years and now understand it clearly. EdTech, unlike most other technology-enabled sectors, is a profoundly skills-intensive domain. In a typical technology company, 12 to 15 skillsets are enough to bring a product to market. In EdTech, that number often exceeds 40.

To build a single learning platform, one must integrate expertise across data science, learning design, instructional design, psychometrics, curriculum architecture, pedagogy, assessment strategy, UX and UI, engineering, behavioral economics, cognitive psychology, AI and machine learning, et al. And that is before you even reach content operations, localization, compliance, accessibility, and student experience. Each component interacts with the other, forming a living ecosystem of ideas, design choices, and algorithms.

This is why the dynamic range between the average and the best in EdTech is unlike anything else I have seen in business. In most walks of life, the difference between the average and the exceptional is modest. The best driver might get you to the airport 30% faster. The best smartphone on the market may outperform a mid-tier one by roughly that same margin. Three-to-one is considered a wide dynamic range in most domains.

In EdTech, especially since the advent of deep technology—AI, machine learning, and generative models—the difference between the average and the best is not three-to-one. It is 25-to-one, sometimes 50-to-one. The best data scientists, psychometricians, and learning scientists I have worked with are not incrementally better than their peers; they are orders of magnitude more capable. Their insights reshape product direction, accelerate learning outcomes, and redefine what is technically and pedagogically possible.

Much of my success as a leader has come from finding these extraordinary people who operate in that upper performance range. I have learned that when you gather enough of them in one environment, something remarkable happens. They thrive on working with peers who challenge them. They set their own bar higher. They begin to recruit only others of their caliber. Excellence becomes self-reinforcing. Over time, these small pockets of A-players coalesce into a culture that polices itself, driven by pride in quality and clarity of purpose.

That has become my central hiring heuristic: always seek people who are significantly better than oneself in their domain. Bring in the exceptional data scientist, the learning designer who understands cognition at a cellular level, the instructional architect who can translate theory into product behavior.

We built such a team: a global, multi-shore, multi-skilled core that became the nucleus of our innovation engine. They came from diverse backgrounds and geographies, united by their craft and curiosity. Together, they created products that stood the test of time because they embodied one principle: never compromise on product efficacy.

The architecture of human–machine collaboration in EdTech is not just about algorithms augmenting teachers or learners. It is about people augmenting one another's strengths through shared purpose and design. The best AI tools are not replacements but multipliers. They elevate what the best human minds can do by removing friction, expanding insight, and amplifying intent.

As leaders, we have a responsibility to be the yardstick of quality. Excellence is not an ambition in education; it is a duty. Human–machine collaboration begins with human mastery, because great systems are built only by people who care enough to make them great. When we anchor the work in human intention, AI becomes a true partner and not a substitute. When great minds create thoughtful tools, the bicycle becomes more than a device; it becomes a vehicle for collective insight, carrying human understanding farther than it could ever travel on its own.

Policy, Equity, and the New Social Contract

AI tools can personalize learning at scale, but personalization without equity risks creating an aristocracy of access. Policymakers must now craft a new social contract, one that safeguards data privacy, algorithmic transparency, and universal access to AI-assisted learning.

True innovation in education policy lies not in regulation alone but in collaboration. Governments must partner with the private sector and EdTech companies to create credential frameworks that recognize AI-mediated competencies while upholding fairness, inclusion, and trust. The future of education will depend not on how fast we deploy AI, but on how wisely we integrate it, ensuring that technology serves as a bridge to opportunity, not a barrier to it.

When learning becomes both equitable and intelligent, we move closer to what education was always meant to be: a collective act of human progress. Equity is the foundation of that progress. AI must serve humanity as a whole, because its power is justified only when it expands access to human potential. A tool is transformative only when everyone can use it. The bicycle changed the world because anyone could ride it. GenAI carries the same responsibility. Its promise is fulfilled only when every learner has the chance to move forward with it.

Scaling Learning Without Losing Soul

In most organizations, if you ask why something is done a certain way, the answer often is, "Because that's the way we've always done it." It sounds harmless, but it is the slow death of innovation. Progress begins when people are encouraged to question the ordinary, to ask *why* not as dissent but as curiosity. The foundation of quality, in my view, lies in treating every process as a living hypothesis. There must be a rationale behind what we do, a clear description of how we do it, and most importantly, a mechanism to question and refine it. That mindset is the true paradigm shift. It transforms a culture from passive repetition to active discovery.

As companies grow, many lose sight of this. They begin to replicate their early success by codifying the processes that once produced it. Over time, they confuse the process with the purpose. The ritual overtakes the reason. I have seen this happen across industries. In technology, giants like IBM and Xerox once faltered under the weight of their own procedure. In education, many organizations that once defined the industry became so focused on scaling systems that they lost sight of the craft: content, pedagogy, and the learner's journey.

In EdTech, the temptation to over-engineer solutions is even stronger. The first ideas we develop are usually complex, layered, and intellectually satisfying. But complexity often masquerades as progress. True mastery lies in finding elegant simplicity: the solution that works not because it dazzles but because it makes sense. That simplicity can only emerge when teams live with the problem long enough to peel back its layers and see its essence. Most leaders do not invest the time or the patience required to reach that clarity.

Scaling learning without losing its soul means holding on to that clarity. It means building systems that are flexible enough to evolve but disciplined enough to maintain purpose. It means resisting the comfort of folklore and embracing the discomfort of inquiry.

Learners today are intelligent, discerning, and evolving. They do not want complicated products; they want considered ones: tools that reflect thought, care, and empathy. As we scale, our challenge is to preserve that sense of craftsmanship. To ensure that process never overshadows purpose and that technology never outgrows humanity.

In the end, great companies, like great schools, are built not on processes but on people who care about doing things well and doing them right. When we scale that kind of care, we do not simply grow, we elevate. The soul of learning is preserved when AI becomes a force multiplier for human insight, because progress in education has always meant elevating people, not processes. As we scale, our task is to keep improving the bicycle rather than worshiping the assembly line that builds it.

The Human Element: Leading, Learning, and Evolving in the Age of Co-Intelligence

The role of the educator has never been static, yet it has never faced transformation at the pace we see today. As technology moves from tool to collaborator, educators are shifting from transmitters of knowledge to architects of learning experiences. They are no longer confined to delivering information; they now orchestrate environments that awaken curiosity and inspire independent thinking. The essence of teaching has become the art of designing meaning.

GenAI accelerates this evolution. It liberates educators from repetitive instructional tasks and gives them back what technology had long taken away: time to teach, time to listen, and time to mentor. In this new landscape, the most valuable educators are those who can blend empathy with analytics, creativity with computation, and pedagogy with data. They do not fear AI's precision; they harness it to amplify their own human touch.

Yet, this transformation is not only technological; it is cultural. Every society brings its own idioms of learning, its rhythms of curiosity, and its relationship with authority and questioning. AI systems must adapt to that diversity rather than flatten it. A curriculum shaped by GenAI must respect cultural nuance as much as it values cognitive rigor. If we ignore context, we risk building systems that are technically brilliant but spiritually hollow. True global education depends on preserving plurality while pursuing progress.

Leadership in this age demands both courage and restraint. The EdTech leader of tomorrow must think like a designer, act like an educator, and decide like a philosopher. Building AI-driven learning systems is not only about data accuracy or product velocity; it is about moral architecture. Leaders must ensure that technology magnifies humanity, not marginalizes it. The most enlightened organizations will be those that build not just intelligent systems, but ethical ecosystems.

Innovation, too, is a deeply human act. It rarely happens in boardrooms or spreadsheets. It happens when five people gather in a hallway at midnight to explore an idea that might just change everything. It happens when someone questions a long-held assumption and invites others to think differently. Innovation is not about budgets or R&D headcount. It is about the people you have, how you lead them, and how deeply they understand the mission. In EdTech, innovation is the act of making existing ideas better, of introducing new methods that simplify rather than complicate.

Learners and instructors often do not know what they truly need until they see it. Our task as innovators is to anticipate those needs. To read what is not yet written in the fine print of education's future.

To this end, two mantras have guided me throughout my career: simplicity and focus. In education technology, simple is often harder than complex. You must work relentlessly to clarify your thinking, to pare away the unnecessary, and to reach the essence of what truly matters. Simplicity in instructional design, visual clarity, and functional elegance requires deep understanding and constant refinement. Focus is equally vital. Innovation often comes from saying no to a thousand things. The discipline to eliminate distractions, to decide what *not* to do, is what keeps a company's vision intact. Sometimes, innovation is about subtraction. A handful of state-of-the-art products built with purpose will always outlast a catalogue of mediocrity.

Over the years, I have also developed an appreciation for intuition, that subtle interplay of experience, instinct, and observation that guides decision-making when data alone cannot. In my view, intuition is not a guess but a synthesis of experiential wisdom built over thousands of design choices and product iterations. It is what allows true innovators to see around corners, to sense opportunity before it fully materializes.

The economics of AI in education depend on that balance between intuition and intelligence, creativity, and computation. Efficiency must never replace empathy. Automation may cut costs, but only human insight can raise quality. Institutions that blend both will thrive. Those that pursue scale without soul will quickly discover that no algorithm can substitute for trust.

Assessment must evolve in the same spirit. Traditional testing often measures memory more than mastery. In a world of generative AI, the true measure of learning lies in interpretation, synthesis, and creativity. Adaptive assessment driven by transparent AI can provide real-time insights into a learner's progress without reducing them to a score. The goal is not to replace teachers with algorithms, but to replace assumption with understanding.

Education itself is no longer a phase; it is a continuum. Lifelong learning has become lifelong co-learning: humans and intelligent systems learning together, each shaping the other's growth. Students will co-create knowledge with AI, asking deeper questions, building connections, and cultivating wisdom in ways we could not have imagined a decade ago.

Leadership, therefore, becomes an act of stewardship. It is about protecting the soul of learning while expanding its reach. The best leaders in this era will not be those who chase every trend, but those who see clearly, focus sharply, and innovate with integrity. They will understand that great technology does not replace people; it reflects the best in them.

In the end, the future of education will not be defined by the smartest systems or the fastest algorithms, but by the wisdom to align innovation with purpose. When we lead with empathy, design with simplicity, focus on what truly matters, and trust our intuition, we create a world where human and artificial intelligence evolve together as partners in learning. That is the essence of leadership in the age of co-intelligence. AI can extend our reach, but only human judgment can determine where that reach should go. Culture shapes the terrain, yet a well-crafted bicycle lifts every learner above the steepest ground. When we design with that level of care, we do more than build technology. We shape the path forward for human progress.

Conclusion: The Next Renaissance

The through-line of this entire chapter comes to a single point here: The future will not be shaped by the intelligence we build, but by the humanity we choose to preserve while building it. We are standing at the threshold of a new educational renaissance. Every few centuries, humanity invents a tool so powerful that it reshapes how we learn, work, and think. The printing press democratized knowledge. The internet globalized it. Generative AI is now personalizing it at scale.

Yet the question before us is not *what* AI can do, but *why* we build it. The true measure of progress will never be the sophistication of our algorithms; it will be the wisdom, empathy, and purpose we bring to their creation. Technology on its own has no soul. It is our intent that gives it meaning, our imagination that gives it direction, and our ethics that give it purpose.

There are many ways to express gratitude for being alive at this remarkable point in time. Some people express it through service, others through art. For me, it has always been through creation. Every EdTech product, every platform, every piece of content crafted with care and integrity is an act of appreciation for the learner, an offering to humanity. We may never meet the millions of students who use what we build. We may never hear their stories or share our own. Yet, in the act of creating something purposeful, something that truly makes learning accessible and meaningful, we transmit an invisible thread of gratitude to the larger human family.

That, I believe, is what it means to build with soul. To honor the privilege of creation by ensuring that what we make helps others rise intellectually, emotionally, and economically. Our responsibility as EdTech leaders is not only to innovate, but to ennoble; not just to design for efficiency, but to design for equity; not merely to scale, but to elevate.

If we succeed, education will move beyond the transfer of information and become an engine of transformation. Every learner will gain a guide. Every educator will gain an ally. Every institution will regain its moral purpose. And if we build wisely, GenAI will become the greatest bicycle ever crafted for the human mind, helping every learner travel farther, think deeper, and rise higher.

About the Author

Russell Ramesh is a market-facing, multi award-winning EdTech thought leader with over 20 years of leadership across digital learning, publishing, and enterprise education. He is recognized for building category-defining products that fuse learning science with frontier technology and for shaping the strategic direction of EdTech organizations through clarity, design discipline, and a relentless focus on learner outcomes. Russell is deeply proficient across the full PDLC of EdTech SaaS, with specialization in architecting GenAI-powered learning systems that personalize instruction, enrich engagement, and scale mastery.

He served as a key member of the Advisory Panel for the US Department of Education's Office of Educational Technology, where his mandate centered on national EdTech policy, digital infrastructure, and the future of AI-enabled learning. Russell has led breakthroughs across robotics, adaptive engines, OER, deep AI and GenAI, data science, and immersive technologies. His work continues to push EdTech toward deeper efficacy, simpler experience design, and meaningful human progress.

LinkedIn: https://www.linkedin.com/in/russell-ramesh-51230510/

CHAPTER 20

HUMANS, MACHINES, AND MONEY: CAN DEFAI REDUCE INCOME INEQUALITY?

By Annette Raynor
Founder of Wealth Engineering; Speaker
Eatontown, New Jersey

The power and advantages of decentralization are becoming increasingly clear. We deserve a financial system where no one can be censored or excluded from full participation.

—Silvio Micali, Founder of Algorand

DeFAI is a new term that refers to the combined capabilities of decentralized finance (DeFi) and artificial intelligence (AI). Application of this technology is in its infancy with only a small portion of the population adapting and using it for financial gain. This will change as we face the largest transfer of wealth in history from traditional finance to a decentralized system.

Before we dive into the details, I think it is important to identify the problem that can be solved. Understanding the severity and impact on society can assist us in overcoming fear and trepidation when it comes to applying DeFAI.

Let's start by looking at some data:

- The World Bank's Global Findex 2025 Report indicates 1.3 billion adults are unbanked, excluding them from our financial system.

- The Federal Reserve indicates 6% of American adults are unbanked as of 2023.

- Unbanked rates are much higher among lowincome adults (23% of those with incomes below $25,000 were unbanked, compared with 1% of those earning $100,000 or more) and are also elevated among younger, Black, Hispanic, women, and disabled adults.

It comes as no surprise that the unbanked tend to be low income and marginalized, after all, they have been left out of our financial system. They lack the ability to complete basic financial transactions. They cannot write a check, use a debit card, apply for a credit card, or any type of loan, for that matter. So, you must wonder how they will ever be able to improve their financial situation if they lack the basic tools of traditional finance.

Decentralized finance is the adaptation and use of digital currency that is not managed by a central authority. It is a set of financial services built on public electronic ledgers that lets people lend, borrow, trade, and save without the involvement of banks or brokers.

When the power of artificial intelligence is applied to this financial ecosystem, we can potentially experience wider utilization. There is no definitive figure for how many Americans use DeFi and estimates range from 6% to 22%. However, nearly 40% of Americans want to learn more about DeFi, suggesting an interest exists but actual participation has not taken place. Most people fear the use due to the lack of oversight and regulatory framework. This is a difficult obstacle to overcome and one that I personally hope we never address.

The foundation of decentralized finance is autonomy. No regulation, no central authority, no government, bank, or financial institution that decides who may participate. DeFi is available to anyone with an internet-enabled smart phone.

The largest security risk to a decentralized network is its connection and interaction with a centralized brokerage. We should not force the traditional financial framework onto the DeFi space. The beauty of DeFi is access and autonomy. I can sell goods and receive payment in seconds anywhere in the world. I can self-store my wealth and stake it when I want to earn additional income. I can put my money to work, and I can decide where and when I want to lend my wealth to the network. It is a difficult concept to grasp, especially when the central financial system is so ingrained in our society.

Artificial intelligence is assisting humans in the adaptation of DeFi. As an example, AI can manage an investment portfolio, and it can do this 24 hours a day, seven days a week. It does not get tired, and it can evaluate more data in a shorter period than any financial analyst ever could. AI can assist humans in finding both borrowing and lending opportunities. AI combined with DeFi can improve capital efficiency by continuously managing risk and reacting faster than human systems could. As an example, AI can automatically move a portfolio to cash in a volatile environment, where it would take time for analysts and institutions to take the same action. AI is also making it easier for us to use decentralization by automating routines for us. AI can find the highest-yielding liquidity pool for your digital currency and can execute multi-step on-chain strategies on your behalf.

At a basic level, DeFAI is considered the next phase after basic DeFi. Instead of just decentralizing the infrastructure, it adds adaptive intelligence on top, with AI agents that can assist us with everything from electronic wallets to network governance.

With DeFAI and a smart phone I can hold my money in digital currency. I can transact with this currency. I can borrow and I can lend without applications or third-party approvals. I can start a business and accept payments from anywhere. I don't need a traditional banking relationship to carry out these activities. I simply need to learn how to use them.

If we know that the unbanked population correlates to lower income and marginalized groups, then we can infer the ability to participate in finance will lead to better incomes and improved cash management. DeFAI expands access to basic financial services and provides lower costs and increased user control, which gives it real potential to narrow income and wealth gaps.

Opportunity

With significant change comes opportunity. As I mentioned, we will experience a significant transfer of wealth from centralized to decentralized systems. Those who are willing to learn, adapt, encourage, and participate stand to benefit significantly.

However, this change comes with significant responsibility. Individuals are taking control of their assets and are, therefore, accepting the risk of management, security, and deployment.

Responsibility

As an example of this responsibility, electronic wallets are controlled by digital keys and secret phrases known only to the owner of the wallet. Wallet owners have lost and misplaced their phrases and have attempted to contact a wallet provider to recover their phrase. This is simply not possible due to the design of decentralized finance. The wallet is made available through software, and the software provider does not have access to the wallet the user created, which is unique to the user. This can be a difficult concept to grasp. There is no customer service or complaint department. You are taking full control of your financial activities, wealth management, and security.

It Is Not One or the Other

As we move forward, it is reasonable to expect that one financial system will not replace the other. Instead, I believe we will utilize both centralized and decentralized mechanisms. The single most important differentiator is access. DeFAI provides the world's 1.6 billion unbanked individuals with access to financial tools and services. This access could represent the ability for us to make a major impact on global income inequality because we are providing lower income individuals an avenue to participate in the world's economy. Individuals do not have to apply, or show an income, or a home address. They can begin to participate immediately. They will have access to the financial tools we take for granted: checking accounts, debit cards, credit cards, personal loans, and lending structures.

Centralized Banking Issues Eliminated

The benefits of DeFAI are not limited to the unbanked. It provides great opportunity for the masses and addresses some of the issues of our centralized banking environment. Currently, we place our money into a bank account that we applied for and were approved. The bank uses our money to make money and provides us with the smallest fraction of this income in the form of interest. While they are using our money for their profit, they are also charging us fees to use their banking services. This doesn't seem fair, but it has been this way for decades.

The bank could mismanage our money; FDIC protection is not guaranteed, especially if the bank doesn't go bankrupt. In this scenario only the bank account holders will lose their money. This is a true statement; you can reference Evolve Bank and Synapse, a fintech provider that has cost individual account holders millions of dollars with no impact to the bank or the fintech provider.

Banks can close our accounts at will, with no explanation and no recourse. Imagine you are a small business and all the money from your sales goes into your bank account. You also pay your vendors and suppliers from this bank account. Then, one day, the bank closes your account with no forewarning. Your business stops. It does not resume until you establish a new banking relationship. I think most people would be shocked to know how common this situation is. A bank can decide whether a business is too risky or an individual does not meet the institution's criteria. These actions stall growth and inhibit wealth creation. No wonder we have income inequality. Our financial systems are designed for those who have means and privilege.

On the other hand, DeFAI allows the individual to custody their money, lend it when and where they want, and have the interest paid directly to them, while also providing them with access to instantaneous loans.

If This Is All So Wonderful, Why Isn't Everyone Using It?

That's a great question, and it can be attributed to power and control. Our advisors, consultants, and financial institutions have considered digital currency a scam, fake, and to be avoided at all costs. Banks have summarily closed bank accounts of individuals who received deposits from crypto exchanges. The traditional financial system has fought long and hard to maintain control, and DeFi represents the loss of that control.

History will show that 2025 was the year our financial institutions finally began to accept DeFi and immediately applied its abilities to traditional financial products. Exchange traded funds (ETFs) enable you to participate in digital currency without the "risk" of participating directly. Once again, this defeats the whole purpose of decentralization and automation, but for those who don't want to learn, these products can be beneficial. You can participate in the transformation without having to learn or assuming greater responsibility.

I think "knowledge" will ultimately control the outcome. Finance has long been the sector that required education, licensing, guidance, advisors, and extensive compliance. An individual could not possibly be expected to understand how to use complex financial instruments for their gain; only a financial professional can do it.

Well, DeFAI changes that. For the first time, "we the people" can learn and take control, making us the specialist and master of our financial destiny.

At the risk of losing your attention, I would like to explain some of the benefits of DeFAI in plain language.

Yield on Idle Assets

Let's say I didn't want to miss out on the crypto craze, and I purchased some bitcoin (BTC) and ether (ETH). I am holding these at a crypto exchange account I set up, and the exchange maintains custody of my coins. I could take these coins into a wallet I created and use an AI agent to direct my coins to the most lucrative liquidity pools (income yield) and rebalance as needed to generate income on the asset I hold.

AI Managed DeFi Lending

You supply your BTC and ETH to lending opportunities that are monitored by AI and readjust your positions in real time to protect your income stream.

Automated Trading and Portfolio Management

Instead of needing professional trading skills, you can describe your goal (for example: "generate conservative income on my BTC and ETH"), and the AI will construct and maintain a diversified on-chain strategy that would be difficult for you to manage solo. Or, machine learning AI can create and manage investment portfolios of traditional financial instruments and monitor and adjust them daily.

Simple Payment and Transfers

Using digital currency, I can send coins anywhere in the world instantly for a few dollars, in some cases a few cents. Conversely, if I wanted to use my bank to send money to my grandmother in Italy, I would pay upwards of a hundred dollars for the wire, and my grandmother would have to wait days before the funds arrive, and she would also pay an incoming wire fee.

Expanded Global Reach

As a business owner I could expand the geography into which I sell my goods and services by simply accepting digital currency for payment. Payment is immediate, irreversible, and inexpensive. As the business owner, I receive payment, and that payment cannot be reversed. I will have to initiate a refund if the client is not satisfied, which is preferable to business owners. This eliminates chargebacks and eradicates credit card fraud. As the consumer, I can pay for goods immediately without incurring any bank charges, wire fees, or the risk of credit card fraud every time I put my card on file. You can't "accidentally" pay someone in crypto currency. You can make a mistake and pay the wrong person, but the payment you make is intentional and cannot be made by someone else. Further, I can significantly increase my sales by being able to sell to a larger global population who have access to digital currency even though they don't have access to a bank account.

Self-Custody

While self-custody doesn't generate money or additional wealth, it does provide security and peace of mind to those who desire it. When you purchase digital currencies and you take them offline into a hardware wallet that you store and protect, you can be sure that your wealth will not be hacked, stolen, lost, or tied up in bankruptcy. In other words, that wealth sits on that hardware that you can put in a safe, a safety deposit box, or under your bed, whatever gives you peace of mind. This is just another example of providing you with total control of your assets.

In summary, we are embarking on an amazing journey. Decentralized finance is possible due to advancements in technology, blockchain, digital currency and artificial intelligence. AI plays a dual role in establishing functions of DeFi but also helps humans to understand and use it for their benefit.

DeFi and AI can seem frightening, overwhelming, and impossible to comprehend, especially when there is so much negativity trying to prevent its widespread use. This is where we must flex our humanity and seek information. By learning and understanding these concepts, you will be able to personally benefit but even more importantly, be a part of shaping the future. As a pioneer, you will help build the roads for generations to follow.

I look forward to what we can accomplish when we embrace technology for the advancement of humankind.

References

1. Leora Klapper, Dorothe Singer, Laura Starita, and Alexandra Norris, The Global Findex Database 2025: Connectivity and Financial Inclusion in the Digital Economy (Washington, DC: World Bank, 2025), https://hdl.handle.net/10986/43438.

2. Board of Governors of the Federal Reserve System, Economic WellBeing of U.S. Households in 2023 (May 21, 2024), https://www.federalreserve.gov/publications/2024-economic-well-being-of-us-households-in-2023-banking-credit.html.

About the Author

Annette Raynor has a varied background with a high concentration in finance and emerging technologies. Throughout her professional career, which includes positions with startups and Fortune 50 companies, she has found that technology applied can improve all aspects of business and personal well-being. Annette is committed to individual investor education, integration of software analysis into all investment strategies, and transparent management practices.

Having an extensive career in emerging technologies since 1983 along with professional financial advisory and trading experience since the early 2000's, a professional educator and an early adopter and deployment of blockchain technology, Annette believes the face of finance will change dramatically over the next 10 years. Her goal is to inform and prepare individuals to take advantage of these changes and prepare themselves to create generational wealth. Annette is an avid investor in options, commodities, stock, private equity, real estate, and digital currencies.

Email: annette@trulyannette.com

Website: https://amped.bio/@TrulyAnnette

CHAPTER 21

EXECUTIVE'S JOURNEY TO OPERATIONAL AI

By Doug Razzano
AI Executive; Co-Founder of Ops2AI
Scottsdale, Arizona

I skate to where the puck is going to be, not to where it has been.
—Wayne Gretzky

AI Is the Puck

The pace of the AI innovation cycle is paralyzing and as fast as a puck on ice. AI is a survival requirement for every company, especially in private equity, advisory, and services-led firms looking to extend capabilities, drive execution and differentiation, and increase portfolio valuations. The future is skating toward AI, and the organizations that don't move with it will get left behind.

Yet most midmarket companies, along with private equity, venture capital, and advisory service firms, don't have technical teams in-house to build AI from scratch. The cost of hiring multiple AI developers can be staggering, throwing off investment theses, blowing through budgets, and ultimately derailing AI adoption. The result is strategic hesitation—and for many, an AI strategy in name only.

Too often, executives are left watching the puck fly across the ice, unsure whether to chase it, pass, or shoot. And by the time a decision is made, the opportunity has already moved.

AI Hype and Delusional Expectations

The AI hype cycle has promised big gains in efficiency, optimization, and automation, but the results have rarely matched the headlines. Most efforts have focused on deploying tools or launching disjointed pilots, with little connection to real workflows or measurable value. The failure rate is high, both in terms of ROI and lasting business impact.

Too many executives fall into the trap of assuming that buying an AI tool equals transformation. But tool deployment alone does not create value. True AI impact comes only when companies adopt an operational AI mindset: embedding AI into a few high-value workflows that are rewired from end to end. This isn't about sprinkling AI on top. It's about redesigning the process around intelligent automation and decisioning.

Operational AI is the real differentiator, and it's measurable. You don't need dashboards full of vanity metrics. You need to show impact on EBITDA, workflow efficiency, or other hard business outcomes. Every technology investment must pay for itself, and AI is no exception.

Executive Hint: *Your AI journey must prove economic outcomes to win.*

From ChatGPT to Operational AI

Operational AI is fundamentally different. It's not about experimenting with tools. It's about embedding intelligence into the actual way work gets done. It means redesigning core workflows so that AI agents are making decisions, automating human tasks, and increasing capacity inside the flow of operations.

Yet according to McKinsey, only 3% of companies have successfully scaled a generative AI use case in an operations-related domain. That tells us something: This is where the opportunity, and the execution gap, lies.

When I first heard the phrase "operational AI," it sounded like a buzzword, a consultant's way to sound uniquely relevant. But as I went deeper, I realized the problem was bigger than terminology. AI tool vendors simply didn't have access to the kind of deep, contextual data AI needs to make smart, repeatable decisions across multiple inference points.

That's when the shift happened.

Every company will eventually build its own internal, GPT-like AI environment. This isn't science fiction. It's an operational necessity. These environments will house a company's intellectual property, tribal knowledge, and nuanced decision-making processes. The real "secret sauce." They will run fleets of AI agents, each responsible for executing tasks, surfacing insights, and protecting corporate assets. They will create living, breathing knowledge bases that evolve with the business.

Executives won't just manage teams. They'll manage AI applications. They'll oversee agent design, deployment, and performance. Human employees and AI agents will collaborate, share intelligence, and jointly execute operational work.

That's the future of AI. And it's already starting.

The Executive Mind Shift to Operational AI

From this moment forward, your ability to speed skate to the AI puck and score is what will define success. The complexity of operational AI demands a new kind of leadership, one grounded in a deep understanding of workflow mechanics, not just strategy decks.

Operational AI starts at the task level. It requires executives to understand what actually happens in the trenches: what tasks produce value, which AI agents can augment or automate those tasks, and how outcomes will be measured. This isn't abstract transformation. It's measurable, designable, and repeatable execution.

Let that soak in.

To succeed, executives must adopt a mindset that blends operational depth with iterative intelligence. The leaders who win will be those who continuously refine and embed AI agents into workflows, leveraging business intelligence, real-time inputs, and performance feedback to guide perpetual learning and improvement. They'll champion a new model of execution: one that is agent-augmented, outcome-measured, and continuously optimized.

But there's a human side, too. The same executives also carry a new burden: defining how AI and people coexist. That means structuring human-in-the-loop governance, building trust across teams, and establishing clear policies on when decisions require human override, ethical review, or cross-functional input.

Executive Hint: *Managing AI agents and hiring people with the skills to do so is coming faster than you realize.*

How Do I Operationalize AI in 2026?

Every technology adoption journey begins with people, process, and technology, but AI raises the bar. It demands a deeper level of technical fluency from every executive and opens the door to a new kind of creativity. This is not about deploying a tool. It's about reimagining your business environment with AI woven into its core.

AI can do nearly anything a human can do. That scale of potential is overwhelming, and it's why many executives feel paralyzed. They don't know where to begin.

Now is your moment. Picture the next ELT meeting. You're asked: "Where will AI have the biggest impact in your part of the business? Where are you starting?"

Executive Hint: *Your only answer is "compliance."*

Why? Because most executives have been dabbling—scattered pilots, vendor tools, pet projects—with no real oversight. But the moment an AI idea gets to the compliance desk, the brakes slam. Proofs of concept die in security reviews, data governance checklists, or lack of documentation.

To escape this trap, you need to reframe your AI approach: Compliance isn't the obstacle. It's the runway. If you can't get past compliance, your AI project never takes flight. But if you start with compliance, you build the foundation for credibility, scale, and business impact.

The next wave of executive-led AI projects will only succeed if they're born with governance in mind. The Wild West era of "move fast and break things" is over. The new mandate is to move fast and validate everything.

Compliance is your path to AI success.

Starting Your Compliant-Ready AI Project

AI is exciting. Every executive has a million ideas about what they want from it. But most have never taken a single AI concept through a formal compliance review or even built something end-to-end with AI. This is unfamiliar territory.

Many companies have project review processes, but what we've observed is mostly unstructured experimentation. Ideas get tossed into vendor tools, dashboards appear, someone demos a "cool feature," and the cycle ends without business validation or governance. That's not sustainable.

The first step toward meaningful progress is building your AI business case inside a standardized AI project template. This template should clearly define:

- The executive sponsor
- The project's working title

- What the solution does
- How it will impact your business domain

Once documented, this clarity helps secure alignment across teams. It forces conversations about ownership, success metrics, risks, and integration points. It also helps identify the right internal partners—security, data, legal, compliance, operations—early in the process, before delays and rework become inevitable.

Executive Hint: *Gaining compliance approval at the start positions your project for proof-of-concept success.*

It's not just about playing defense. It's how you give your AI project a real chance to scale, create value, and survive the scrutiny of your ELT, board, and regulators.

AI cannot be a free-for-all anymore. A governed starting point, built with a clear template, upfront alignment, and compliance-ready framing, is what separates toy pilots from production-ready solutions.

AI Gotcha Gaps

Like most of my journey into operational AI, it's been anything but clean or structured. Messy is a better word: wrong turns, false starts, and a lot of underestimations. I've seen firsthand the gaps in executive understanding, the emotional hesitancy, and the fear that comes with this much change. It's not just new technology; it's a full-blown behavioral shift.

Honestly, it's felt a lot like the gold rush of 1849. Go West. Avoid disease. Buy a shovel. Start digging. Some struck gold, but most didn't. The point is, this chapter isn't a polished success story. It's a field report. Leaders don't need more hype. They need the truth about what it takes to bring AI into the core of a business. Here's my attempt to give it to you straight.

Your Team and Leadership

Your biggest obstacle isn't the tech. It's the people. AI gets personal fast. People protect their roles, their relevance, their routines. If you're not managing the emotional side of adoption, the technical side won't matter. Resistance will win.

Given how fast AI is changing, people crave clarity. They want to understand how AI affects their roles, why it matters, and where they fit. AI agents will inevitably take over some of your team's daily tasks. That fear is real. Ignore it, and you lose momentum before you begin.

The Data

AI exposes the ugly truth about your data. And it's not pretty. Operational AI shines a spotlight on things your company has ignored for years: disjointed systems, inconsistent CRM entries, duplicate customer records, untraceable financials, and tribal knowledge locked in the heads of two legacy employees.

As an executive, you're often downstream of the data, reliant on others. But if you want operational AI to work, you need to own the state of your data. You don't need perfect data, but you do need clean, accessible, and consistently structured data.

The Broken PoC Model

To limit the failure rate of AI adoption, executives must recognize and address the blind spots baked into the current PoC (proof of concept) model. Most PoCs fail. Not because the idea was wrong, but because the conditions for success were never established in the first place. Here's what typically goes wrong:

- The data isn't ready, or it's of poor quality.
- There's no clearly defined ROI or outcome.

- Governance and compliance are either missing or come in too late.

- User adoption is overlooked or assumed.

- There's no built-in runway to scale beyond the test phase.

Executives enter PoCs full of hope, but without the infrastructure to move from concept to production, the idea dies in the middle. Worse, they often don't know why.

To succeed in operational AI, your organization needs a new framework, one that intentionally de-risks these predictable points of failure. That starts with a clear-eyed assessment of your company's AI readiness. Are your teams aligned? Is the data clean enough? Are your compliance and governance partners involved from the beginning?

Without these, even the best AI idea becomes just another cautionary tale.

The Company AI Lab

The documented failures in the rush to capture AI value are beginning to change. Between 2023 and 2025, most companies failed to cross the threshold into true Operational AI. Why? Because they lacked the guardrails: compliance checks, robust testing protocols, agent engineering standards, and ongoing data hygiene practices. The result was chaos: unstructured AI experiments launched in isolation, with no unified governance or ability to scale.

Today, companies need an intentional launchpad. Not a vendor-led experiment. Not a rogue pilot. A structured, internal environment where every AI idea can be stress-tested, tracked, and matured.

This is the company AI lab. Think of it as a governed sandbox. Every AI initiative begins here with an executive-initiated application registered in a centralized system. That registry triggers a standard compliance approval process and documents the use case, data access needs, risk assessments, target users, and measurable outcomes.

Once approved, the lab manages the entire lifecycle of the PoC. From agent design and user testing to auditing, fine-tuning, and deployment, everything is tracked in one place. The lab becomes the living archive of your AI development history, continuously learning from each new attempt.

But it doesn't stop there. The company AI lab also becomes the home of:

- Standardized prompt libraries and training materials
- Enforcement of policies and role-based access controls
- Performance monitoring of AI agents in production
- A governance dashboard that spans the full portfolio of AI projects

Instead of reinventing the wheel for every new AI use case, your business now has a perpetual, reusable operating environment, one that enables safer, faster, and more disciplined AI adoption at scale.

The Operational AI Executive

The executives who win the next decade won't be the ones trying the flashiest tools. They'll be the ones who transform AI from scattered ideas into a disciplined, repeatable operating system. This is the minimum requirement for leading in the era of Operational AI.

Instead of asking, "Which AI tools should we try?" the real questions become:

- What can AI do at the workflow unit level?
- How do we measure each AI agent's contribution to business value?
- Is there a standard path from compliance to PoC to validation to production?

- Can we scale and govern AI without creating operational bottlenecks?

- Are our AI projects getting safer, faster, and more repeatable with each iteration?

These questions define the mindset of an operational AI executive. They reflect not just technical leadership, but credible leadership, what boards, regulators, customers, and employees now demand. AI is powerful, complex, and transformative. It must be explainable. It must be controlled. And it must be measured.

That kind of transformation doesn't happen by accident. It happens when executives insist on a lifecycle model for AI adoption, one with centralized registration, clear guardrails, auditable processes, and continuous refinement. A system, not a series of experiments.

Adopting these principles doesn't just improve outcomes. It changes how you lead.

You don't need to be the deepest technical expert in the room. But you *do* need to be the one who insists that every AI project starts with compliance, includes clear performance metrics, has defined team roles, and delivers visible business value.

You shift your leadership posture from AI enthusiast to AI strategist.

You push your teams to stop launching one-off pilots and start building inside governed, reusable environments. You treat validation as mandatory, not optional. You demand adoption metrics before declaring success.

AI change management isn't just a technical problem. It's a leadership practice. One that's new for nearly every executive.

Production means accountability, ongoing monitoring, and operational rigor. And when you get it right, AI moves from a flashy experiment into the core of your business strategy, operations, and culture.

Your teams become more productive. Your employees feel safer, because the boundaries are clear. Internal risk and compliance partners stop being blockers and start becoming champions. Your board supports your every initiative. And your own career trajectory accelerates because your AI projects show measurable impact on revenue, cost, and execution.

The choice is clear.

Executive Hint: *Drive your organization to seek AI partners or acquire the talent needed to build your AI Lab and operational runway. If you don't, your AI journey will keep stumbling and fall short of true operational impact.*

Because in this game, the puck is still on the ice, fast, relentless, and moving toward the future. You either learn to skate where it's going or risk being left behind.

About the Author

Doug Razzano is a seasoned leader with 30 years of entrepreneurial drive and a proven track record in early-stage tech startups and strategic leadership roles. Through his pragmatic, value-driven approach, Doug empowers teams to seize business opportunities and deliver results. His experience spans building global ecosystems and driving innovative solutions to market with industry giants like Accenture and IBM Global Services.

Doug is the founder of ScaleOut Partners, a global executive search firm accelerating over 150 Series A-C technology startups from SVPs and GTM teams to exit events.

Notably, Doug excels at identifying high-value opportunities and starting in 2023 AI being his latest endeavor as a co-founder of Ops2AI. Ops2AI is an AI development and AI-managed services. Doug thrives on scaling technology ventures and delivering value to his clients.

Email: doug@ops2ai.com
Website: www.ops2ai.com
LinkedIn: https://www.linkedin.com/in/dougrazzano

CHAPTER 22

SENSE. SYNC. SIMULATE.

By Anthony Rizk, PhD
Innovation Lead, Co-Founder @ idealworks
Munich, Germany

Science may be an incremental pursuit, but its progress is punctuated by sudden moments of seismic inflection.

—Dr. Fei-Fei Li

A New Industrial Era Is Around the Corner

On a cold November day in 2022, the world witnessed another sudden moment of seismic inflection. With the launch of OpenAI's ChatGPT, few could have anticipated the far-reaching and future impact it would have. Until that point, artificial intelligence (AI) was largely seen as a specialized tool, only accessible to programmers, coders, and researchers who worked with it at a technical level. However, this moment in November marked the democratization of AI, putting advanced technology and reasoning intelligence in the palm of everyone. This democratization of AI did not just make technology more accessible; it also paved the way for new concepts like "physical AI." The latter refers to artificial intelligence systems that can sense, interpret, and act in the physical world, such as robots equipped with AI.

This led to bridging the gap between digital intelligence and real-world applications. As these ideas began to grow, it became clear that Industry 4.0, the era defined by big data, the internet of things, and automation, had its limitations. Its siloed, fragmented solutions often neglected the importance of human involvement and collaboration. This shift highlighted the need to move beyond isolated approaches and embrace a new paradigm: robotics ecosystems. In this context, an ecosystem means creating interconnected and collaborative intelligence, where technology and people work together seamlessly. This transition sets the stage for a more human-centric future. It's the stage for the next revolution: Industry 5.0

Industry 5.0 is built on four main pillars: advanced simulation technologies, centralized and connected decision-making systems (such as fleet management platforms), robotics hardware, and software-defined solutions powered by artificial intelligence. At its core, these four pillars revolve around creating a symbiotic relationship between human creativity and AI innovation. As we enter this era, smart factories will no longer be defined solely by automation, but by intelligent ecosystems that learn, adapt, and co-create alongside their human counterparts. Rather than focusing on isolated technologies, the emphasis is on integrated robotics ecosystems. Key components of this revolution include (1) digital twins serving as dynamic, living blueprints of industrial operations; (2) AI as foundational infrastructure instead of a supplementary good-to-have feature; and (3) the next generation of connectivity through 5G/6G. In the following sections, we will explore these three key components and highlight their roles in shaping the future and reinventing robotics.

Digital Twins: A Live Instance of the Factory of the Future

From a definition point of view, a digital twin (DT) is a virtual representation of a physical object, system, or process. The level of realism can vary depending on what is needed, from basic process emulation to highly detailed replicas of the real world with physics laws, such as light reflections. Such realism capabilities have been made possible by advancement in graphics processing units (GPUs) and powerful simulation engines like NVIDIA Omniverse. Furthermore, the latest advancement in digital twins (DTs) allow for the ingestion of real-time data from sensors and internet of things (IoT) devices. This supports hardware-in-the-loop integration, which means that real hardware components are connected to the digital simulation, allowing for real-time interaction and testing between virtual and physical systems. As a result, DTs can simulate real-world scenarios, enable remote monitoring for physical systems, and facilitate prototyping and collaboration through shared, interactive environments.

In the context of physical AI and robotics, DTs address a key challenge: AI systems, especially those controlling robots, require extensive training and testing. However, doing so in the real world can be expensive and risky. Digital twins and synthetic data provide a solution by creating safe, low-cost environments for training and testing. For instance, automotive manufacturers could use DTs to simulate entire assembly lines, allowing engineers to optimize robot coordination and process flows before any physical changes are made to the factory floor.

DTs are evolving from being simple replicas or testing tools to becoming strategic drivers of innovation. In robotics, DTs unlock a new paradigm for accelerating testing and innovation cycles while reducing the cost of trial-and-error loops, a critical benefit in high-stakes industrial environments such as factories or production lines. Moreover, as robotics fleets expand, DTs can scale in complexity. Robots are typically deployed in fleets rather than individually, which give rise to complications and the need for coordinated behavior within the fleet and with other systems. DTs offer an additional layer for optimization, to refine routing planners, task allocation, monitoring, mission assignment, smart device integration, and other fleet-critical functions that are nearly impossible to test at full scale immediately in real life.

At the Industry 5.0 frontier, DTs serve as collaborative hubs for engineers, operators, data scientists, designers, assembly line workers, and anyone who interacts or works alongside robots. They foster a culture of collaboration between humans and machines, allowing all stakeholders to immerse themselves in a preview of the future and envision how upcoming changes will affect their work. This reduces uncertainties and builds trust in technological transformation, helping to keep humans at the center of the digital transformation and amplifying their judgment and creativity.

AI as Infrastructure: From Mere Feature to Foundation

Building on this collaborative foundation, the next evolution resides in the integration of AI as core infrastructure. No longer just a feature, AI crossed a cultural and practical threshold from specialist tool to broadly accessible, consumer-grade product. That shift matters because accessibility changes who can experiment, who can innovate, and how quickly ideas turn into products. Generative models now assist with code, documentation, design variants, and simulation assets; multimodal systems parse images, sensor streams, and text; and increasingly capable agents coordinate tasks across software and hardware. On the robotics side, advances in perception, foundation models for manipulation, and reinforcement learning from human feedback are shrinking the time from "first grasp" to reliable autonomy. Together, these developments transform AI from a narrow accelerator into a general-purpose collaborator.

The latest AI models include multimodal systems, which process and interpret multiple types of data such as files, images, text, and sound. The result: Robots acquire the ability to understand and interact with dynamic and complex environments for enhanced functionality. Through this shift into reasoning, robots evolved from simple automated task executors to a human collaborator, learning from feedback and adapting to new situations. This is unlocking the true potential of physical AI, where robots stand to become part of the infrastructure that scales human ingenuity rather than replacing it.

When ChatGPT was released in 2022, I initially thought (young and naïve I was) that we had reached the pinnacle of AI innovation, almost as if the field had been "solved." I couldn't have been more wrong. Working in robotics, especially when deploying AI outside of the purely digital domain, introduces layers of complexity that many underestimate. One of the most pressing issues ahead is sustainability. Today's rapid advances in artificial intelligence are fueled by an ever-increasing demand for computation, energy, and resources. The success of AI is driving this relentless pursuit of scale, often overshadowing the hidden costs that come with such progress. The current paradigm in AI development is driven by scaling, larger models, more data, and bigger supercomputers. While this approach yielded remarkable results, the next generation of AI would require a shift towards what could be called resource-aware AI.

At the same time, high-quality human data is projected to be fully put to use within the next decade, driving the field towards synthetic data generation and the need for DTs. These needs are highlighted by the transformation undergone by AI from tool to a foundational part of the infrastructure powering industries. This redefines the problem as a system design one. The drive for scale can mask underlying weaknesses, as models trained on web-scale datasets still generalize poorly outside their training distribution and remain surprisingly fragile to noise. This reflects a broader limitation of statistical learning that lacks true cause-and-effect reasoning to produce causal world models that humans naturally use to understand the physical world around them. Such reasoning is especially crucial for physical AI, systems that must operate reliably in real-world environments.

Neuro-inspired architectures, such as Liquid Neural Networks, Recurrent Visual Processing, and Predictive Coding, could target these gaps. These approaches leverage sparsity (using fewer computations), recurrence (repeated processing of information), and causality (understanding cause and effect), enabling richer and more efficient computation. The future of AI won't be determined by scale alone. Instead, progress will depend on treating AI as foundational infrastructure, one that is designed with resource-awareness at its core.

This means rethinking the underlying principles that drive sustainable, robust, and adaptable systems. Alongside the technical challenges, there are important organizational challenges to address. The adoption of advanced AI will require workforce adaptation through AI literacy, ethical consideration, and careful management of integration costs. Overcoming workforce resistance and ensuring equitable access to new technologies will require targeted training programs and inclusive design practices. Similarly, fostering interdisciplinary collaboration and investing in robust cybersecurity measures will help build trust and resilience within these evolving systems.

Ultimately, by embracing technology with mindfulness, we can move beyond building ever-larger AI systems and focus on creating solutions that are fundamentally smarter, more sustainable, and truly human-centered.

The Future Is Connected: 5G/6G Role in the Symbiosis

When we talk about an ecosystem in robotics and AI, we are referring to a network of interconnected elements, working seamlessly together. As previously stated, the future of robotics will move beyond silos and isolated deployments toward a holistic ecosystem that includes robots, intelligence on edge, fleet management capabilities, and digital twins. For this vision to become reality, it's essential to connect these components both physically and intelligently.

On an intelligence level, what connects components is information, which is data. This ecosystem generates a massive flow of data and information between all its elements, aiming for real-time efficiency, decision-making, and responsiveness. However, the evolution of robotics and AI in industrial environments faces challenges due to the limitation of current Wi-Fi technologies. The latter often struggles to deliver the ultra-low latency, high reliability, and real-time data transfer needed for safety-critical physical AI deployment. These challenges are intensified by the massive data volume produced by sensors, cameras, and lidars, all crucial to provide perception for all types of robots. To manage this data, some computation is typically offloaded from edge devices on the cloud, but this process is currently bottlenecked by wireless limitations.

5G and later 6G technology are poised to revolutionize the creation of truly connected industrial ecosystems by linking embodied AI, cloud-based AI, fleet managers, and digital twins. This connectivity helps close the intelligence cycle: Embodied AI offers immediate analysis and response at the device level, while cloud-based AI and fleet managers provide centralized intelligence and broader view across operations, processes, and connected devices. For instance, advanced robots will be able to synchronize with digital twins for rapid prototyping and simulation, while fleet managers optimize resources and workflows across distributed sites in real time.

Building the Future, One Step at a Time

Digital twin, AI ecosystems, and 6G may sound like science fiction and often appear exclusive to tech giants. However, the future of these technologies hinges on their widespread, practical adoption: the use cases. Because real value is created at the level of concrete use cases, this is where the market will develop and where AI, robotics, and related technologies will be monetized and scaled. It is at the use-case layer that these technologies gain a clear business rationale and a reason to exist, whether in healthcare, logistics, production, or even at our own homes doing our daily chores.

Despite the numerous fields that could be disrupted by these technologies, the path to adopting them is rarely straightforward. One major barrier is the gap between cutting-edge research and the realities of day-to-day industrial practices. Researchers often operate free from the constraints of compliance, legacy equipment, and the complexity of scaling solutions in production. In contrast, organizations must grapple with regulatory requirements, ISO certification, aging infrastructure, and considerable integration debt that comes from merging new technologies into established workflows. Workflows that cannot be interrupted.

To overcome these challenges, organizations should begin with the customer and the use case, not the technology itself. By starting with a clear pain point, such as improving quality inspection or material handling, teams can map out where AI and robotics could assist. Standardizing data and model pipelines, investing in continuous validation, and implementing version control for data, models, and code are critical steps for ensuring rapid iterations and integration.

Necessary to the success of this transformation is investing in people. Contrary to the common belief that AI will eliminate jobs, these technologies frequently create new roles and opportunities, provided the people in the organization are equipped with the necessary skills. As we enter the next industrial revolution, organizations must prioritize strengthening domain expertise and enabling people to adopt robots, AI, and DTs as new tools to unleash their creativity. These technologies amplify human capabilities. Developers can accelerate coding with AI, while robotics enhances efficiency and reduces physical strain on the assembly line.

Yet such transformations require more than mere technology deployment, they demand a human-centric approach where employees co-create the future workflows rather than having them imposed. This collaborative model fosters adoption, innovation, and trust, making change management a critical enabler for unlocking the potential of human-machine symbiosis.

Ultimately, the promise is to push the boundary of what is possible: AI and robotics are the means; disciplined adoption is the method; elevated work and better products are the outcome. This opens the door for a connected ecosystem that can sense, sync, and simulate.

About the Author

Anthony Rizk is a technology innovator and co-founder at idealworks operating at the intersection of artificial intelligence, robotics, and digital twins, to make Physical AI a reality. With a career rooted in solving complex real-world challenges, Anthony specializes in architecting intelligent systems pushing the boundaries of how robots understand, navigate, and collaborate in physical environments. Driven by an ability to bridge cutting-edge research with practical, scalable deployments, Anthony has led initiatives to leverage AI-powered solutions in creating next-generation robotics ecosystems that synchronize the physical and digital worlds.

Through work rooted in computer vision, generative AI, such as diffusion models and visual language models, and NeRF-based scene reconstruction combined with his domain expertise in robotics, he blends technical depth with product vision. He has built and guided teams that deliver robust industrial solutions to champion a future where robotics ecosystems are safe, adaptive, and continuously learning.

Website: www.antrizk.com

LinkedIn: https://www.linkedin.com/in/antrizk/

X (Twitter): antrizk

CHAPTER 23

FROM CONTROL SYSTEMS TO SOCIAL SYSTEMS: RETHINKING AI'S ROLE

By Christian Schenk, PhD
Development Engineer, Respect-Ambassador
Lohr am Main, Germany

We are drowning in information, while starving for wisdom.
—Edward O. Wilson

Nowadays, we retrieve more information than our grandparents ever could—yet we still struggle to find meaning.

Thanks to artificial intelligence (AI), we collect, process, and retrieve information faster and more efficiently than at any point in human history. Smart devices and high-performance computing promise comfort, ease, and productivity, while AI systems take on tasks we would never have entrusted to machines just a few years ago. AI transforms the way we work and communicate, accelerates progress, and decrypts complex correlations across science, industry, and daily life. At the same time, society already struggles to cope with the sheer volume of information flooding our attention. AI does not merely add to this stream; it amplifies it.

Since AI's breakthrough and rapid adoption in both industry and private life, an essential debate has emerged: What is AI truly capable of doing for us, and where do its limits lie? Which expectations are grounded in reality, and which drift into illusion or science fiction? More importantly, how should AI be used in ways that strengthen human capability rather than diminish human judgment?

This essay explores these questions in two connected dimensions. First, it examines AI from a technical and industrial perspective, drawing on real-world engineering constraints to challenge the assumption that more complex AI is always better. Not every system benefits from large models, unlimited computation, or full automation, and understanding where restraint is necessary is as important as understanding what is possible. Second, the essay shifts to the social dimension, asking how AI reshapes human interaction, responsibility, respect, and trust in an age where reality itself can be simulated.

With my background as a control engineer and respect trainer, I approach AI not only as a technological tool, but as a mirror that reflects how we think, decide, and relate to one another. My aim is to advocate for AI that serves people—not the other way around—and to show that technical judgment and human-centered values are not competing goals, but deeply connected ones.

Fears and Expectations

Seven years ago, in 2018, Google AI's release of BERT marked the beginning of large language models (LLMs) as we know them today. This breakthrough launched the development of multiple agentic AI systems, all of which have since sparked both expectations and fears. We expect AI to democratize access to information and knowledge, yet worry about losing our ability to think critically. We are increasingly interacting with AI, often at the expense of friends, colleagues, and family. We expect these AI systems to be highly efficient, yet we fear that very efficiency, wondering if they will take our jobs. We demand AI to be embedded in every device, believing "AI can do everything" while ignoring the immense effort and consequences of training and deployment.

When a tech company's executive board mandates the integration of AI into all of its company's products, three typical reactions arise. The first and most dangerous is the "Yes, we can!" mentality. This approach often leads to a kamikaze-like charge across a minefield of unspoken limitations and unclear requirements. The result? Six to 12 months of endless clarification meetings, workshops, and ultimately a phase of disillusionment, all at the cost of wasted salaries and time, simply due to a lack of impulse control.

The second reaction is "Back down." When everyone retreats, you hope you're not the only one left standing. This response may burn some money, but it's usually recoverable as a middle ground between the first and third reactions.

The third reaction is the "Let's clarify conditions and expectations first" approach. Unlike the others, this perspective neither dismisses AI's potential nor ignores its challenges. It focuses on possibilities rather than wishful thinking, acknowledging both opportunities and risks. No one wants to waste time or money, so the logical first step is to ask: What do we expect from AI? What are the real requirements and limitations?

In society, the term AI still conjures images of LLMs, retrieval-augmented generators (RAGs), large contextual models (LCMs), and large behavioral models (LBMs) embodied in humanoid robots or a ghost inside of a machine. We imagine intelligent algorithms outperforming classical methods in every scenario: highly accurate predictions, predictive maintenance, and seamless device communication. In short, we envision high-performance solutions. The cost: an insatiable appetite for data and energy. These systems demand multi-core parallel computing and substantial amounts of memory to function because of the incredibly high number of parameters, surpassing 600 billion parameters to be tuned. This is how society imagines state of the art technology.

Let me draw a picture of what state of the art really looks like. My expertise lies in control engineering. Nowadays, I test, deploy, and develop alternative control techniques for high-performance industrial hydraulic devices, such as proportional valves. Equipped with digital onboard electronics, these products guarantee the highest standards in convergence time, disturbance rejection, and trajectory tracking. For example, valves reach a desired reference point in less than five milliseconds—from 0 to 100% of their maximum range of motion. That's faster than an eye blink and comparable to the wing flap of a hummingbird during mating season (2.5 to 10 milliseconds). These devices operate reliably under extreme conditions: high temperature gradients, contaminated oil, and intense pressure. Factor in inductivity and magnetic fields, and this performance pushes physical limits, making it a masterpiece of control engineering. Imagine driving a tiny nail into a wall to a precise depth—with a sledgehammer, at high speed. That's the level of precision these valves achieve.

Such performance demands every available resource; even minor additional computational loads can be critical. "Some might argue, Just add more computational power" Every price increase risks alienating customers and benefiting competitors.

In conclusion: To fulfill the promise of highly efficient, high-performing, robust, and reliable products, companies must drop the idea of complex AI methods, such as LLMs or SLMs, on onboard electronics.

Does this mean we should abandon the idea of AI? By no means.

Here emerge two thoughts of feasible applications of AI in technical systems:

1. Downscaled small neural networks, for instance, can assist. We have to distinguish AI as the main component acting underneath and AI as a feature or assistant supporting the main component of a system.

2. Everything we now call AI has existed before in slightly different forms; we simply didn't label it as AI.

Let's explore what's inside this toolbox of possibilities.

Artificial neural networks (ANNs), a subset of AI, mimic how neurons in the human brain transfer information via electrical potentials. Each neuron in an ANN receives weighted input from its predecessors, applies an activation function and offset, and produces an output. This process is simple, too simple, in fact, to handle complex mathematical tasks on its own. A bunch of neurons is hardly a paragon of intelligence compared to the apparently intelligent billions of parameters of big LLMs. However, their simplicity and operational principles make ANNs attractive for control applications. Besides, closer inspection reveals that many ANNs are essentially inflated versions of classical approaches, such as sliding-mode control, fuzzy systems, or PID controllers. With some effort you can transform the mathematical equations describing a classical controller to the structure of an ANN. That's the disillusioning truth.

Given customer requirements and computational constraints, highly efficient specialized products like industrial valves offer little room for integrating AI, especially large language models. The true intelligence are the developers. Those employees and pioneers who find intelligent solutions to challenging problems. Problems which require the ability to extrapolate and transfer solutions on a meta level to other problems in a short time without additional training. These employees are a company's most valuable asset.

Does that require training in artificial intelligence, large language models etc.? Definitely.

I am convinced that the need to incorporate AI in our daily work is inevitable even though AI cannot gain acceptance in every nuance of a product.

All these thoughts boil down to one statement: If companies want to bring in the harvest from the field of AI to the table to satisfy the hunger for innovation and sustainability, those companies have to invest in employees. Instead of focusing on highly complex and expensive AI solutions on every level, companies have to empower employees to use AI to gain space for personal development and social interaction. All these thoughts lead to the question of how we handle AI in a social context.

The technical conclusions drawn from industrial systems point to a broader truth that extends beyond engineering. The decision to limit AI, to use it selectively as an assistant rather than a replacement, is not merely a matter of computational constraints or cost efficiency. It is a matter of judgment. The same discernment required to design robust technical systems is required to navigate AI's impact on society. How we deploy AI in machines reflects how we value humans in social systems: whether we prioritize efficiency over understanding, automation over responsibility, and output over meaning. To understand AI's role fully, we must, therefore, move beyond hardware, models, and control loops and examine how AI reshapes human interaction, trust, respect, and our ability to distinguish reality from illusion in everyday life.

Living in a World with AI

Renowned Taiwanese-American computer scientist, venture capitalist, and author, famous for his leadership in AI, Kai-Fu Le observed, "If AI ever allows us to truly understand ourselves, it will not be because these algorithms captured the mechanical essence of the human mind. It will be because they liberated us to forget about optimizations and to instead focus on what truly makes us human: loving and being loved."

This profound statement by Kai-Fu Lee envisions a prosperous, human-centric AI future. I would like to frame this in the context of respect and add another statement by René Borbonus, a renowned speaker, author, and expert: *"Respekt ist Liebe in Zivil,"* which translates as "Respect is love in civilian clothes." Borbonus's words define respect as love.

Combining both statements, we derive what truly makes us human: respecting and being respected.

At the same time, this vision is contrasted by the following warnings from Nobel-winning economist Joseph Stiglitz: "Unregulated AI will worsen inequality" and "The IMF has expressed 'profound concerns' about massive labor disruptions and rising inequality. AI could lead to job losses not only in lower-skilled but also in higher-skilled roles. Generative AI raises profound concerns about massive labor disruptions and rising inequality."

The debate about the risks and opportunities for our society has skyrocketed, leading to an eruption of socially relevant questions: Will AI take my job? How do we deal with AI-generated fake news and reality? Will we use AI to build a world that values people, or will we let it build a world that values only efficiency? Answering these questions will help us find our place in this AI-framed new world, respecting our nature as human beings.

If well considered, all these aspects contribute to a discourse nourished by respect and accompanied by a focus on solutions. If we want answers to these questions, we must stand up for respectful discussions and debates.

Finger-pointing, however, stifles the search for solutions. Discussions in which AI skeptics accuse innovators and AI enthusiasts of being naïve and innovators accuse skeptics of blocking human progress are poison to the process of finding a respectful way of living with AI.

The solution: Instead of nitpicking at the vices we perceive, we need to acknowledge the virtue behind someone's position. Assuming everyone expects the other side to have the same information, background, and point of view, we start discussions on completely different levels. However, consensus requires clarity and openness, paired with the will to understand the person in front of us. The person who is afraid of losing their job may not be afraid of the job loss itself. There may be the responsibility to support a family and the fear of being unable to feed their children. It's about basic needs, responsibility, and meaning. When we want to make a difference, we must understand this first. We must understand the person in front of us. Respect means seeing the other person's needs, fears, and motives. Respect means being on par with the person in front of us.

Both respect and clarity are inseparably connected. Clarity without respect is cold and leads to entrenched positions. Lack of respect and clarity generalizes, simplifies, and creates an atmosphere of despair. Respect without clarity, however, must lead to active listening. And active listening leads to clarity, focus on solutions, and answers.

But what if AI blurs the line between AI-created content and reality? What if fake news and reality look almost the same? In September 2025, a video of an orca attack in a water park flooded social media like a tsunami. In this video, you see an orca during a show accidentally inflicting life-threatening injuries on its trainer, Jessica Redcliff. Right after this horrible accident, the water inside the pool turns red, people are screaming, and the pain-stricken face of Jessica Redcliff makes us realize what our mind doesn't want to believe. Her face, as she is pulled out of the water, is a masquerade of pain and suffering.

The truth: It's fake. There was never such an orca attack, and what seemed to be a documentation of a living nightmare never happened. But this is not the worst. In the presence of that (fake) news our neo-cortex shut down in eco-mode. We cannot think clearly anymore and rely on basic functions and routines only. Our ability to distinguish real from fake news is reduced.

What if, in a realistic video, the president of one nation declares war on another? What if there is a video announcing a worldwide stock market crash? This kind of content fuels global conflicts, and what starts with a virtual video affects the real world tremendously.

How can we distinguish between fake AI content and real content? Today in 2025, we can still unmask AI content using several criteria: AI-created videos still have an unrealistic, error-prone texture, sometimes looking just too perfect. This applies to the surface of a face as well as to water and ground surfaces. Another indicator is voice synchronization: usually, voice and facial expressions do not match.

But what if AI fills these gaps and we can no longer distinguish fake from reality? Even then, we are not defenseless. Here are some tools to help you stay in charge:

1. Be skeptical and sensitive. To be skeptical means to challenge statements and also our perception. In the case of bad news, search for counterarguments and positive examples. Is the source reliable? The same holds true for overwhelmingly good news.

2. Even more important: Pause and center yourself in the presence of astonishing news and content. I received this advice from a respiratory-trained firefighter when I served as a volunteer at the federal agency for technical relief. In critical situations, this advice might save your life. In the presence of fake news, it might keep your thoughts clear and yourself focused.

Another very powerful quote is from Viktor Frankl, who survived the Holocaust during the Nazi regime as a prisoner in a concentration camp. Despite what he experienced, he said, "Between stimulus and response there is a space. In that space is our power to choose our response. In our response lies our growth and our freedom." If you want, you can pause here, focus on this powerful statement, and reflect on what it might mean for your life. Being aware of our reactions, options, and inner state strengthens our robustness, resilience, and ability to make reasonable decisions.

Finally, I want to close this section with one important remark: The more AI mimics human behavior, and the more we think of replacing real people with AI, the more we will thirst for real social interactions with real human beings. There are AI-based, very advanced learning platforms, classes, and courses that seem to make teachers and trainers obsolete. However, there are some things AI cannot replace: (1) the ability to extrapolate and adapt to unknown situations and scenarios; (2) authentic connections between two human beings; and (3) the perceived presence of a real person. Replacing human beings with AI compares to exclusively consuming preprocessed, synthetic junk food instead of healthy, fresh, organic food. There will always be something missing. And this small but important piece makes the difference between sick and healthy, hollow and fulfilled, just existing and truly living.

To conclude: Artificial intelligence will continue to expand what is technically possible, but it cannot replace human judgment, responsibility, or meaning. In technical systems, AI is most powerful when it supports human expertise within clear constraints rather than attempting to override it; in social systems, the same principle applies. When AI is treated as an assistant instead of an authority, it creates space for clarity, respect, and thoughtful decision-making, both in engineering and in everyday life. The future of AI should therefore not be measured by how closely machines imitate humans, but by how well they help humans remain human: capable of understanding limits, distinguishing truth from illusion, and engaging with one another in ways that preserve dignity, responsibility, and purpose.

About the Author

Christian Schenk, born in southwest Germany, started his academic journey in mechanical engineering at the University of Stuttgart, where he focused on thermodynamics and control theory. This foundation laid the groundwork for his deep understanding of systems that would later influence his work with artificial intelligence. Following his studies, he spent four formative years at the Max-Planck Institute for Biological Cybernetics in Tübingen, immersing himself in research before earning his doctorate degree in control theory. His transition from academia to industry led him to Bosch Rexroth, where he applied his expertise as a control engineer, particularly in developing alternative strategies that included AI-based methods. This period was crucial in shaping his technical acumen and its practical application.

The onset of the Corona Pandemic prompted him to reflect deeply on two pivotal questions: What makes communication successful and how can we maintain respectful interactions in discussions and debates in different social settings? Four years later, in

2023, he started actively to foster respect as a cornerstone of human interaction. His commitment extends beyond individual interactions; he believes in creating environments where respect nurtures personal growth and collective progress. Today, he conducts seminars and workshops focusing on personal development, emphasizing the importance of respectful mindsets, especially among leaders.

He advocates for an approach to AI that prioritizes humanity. His work in AI is not just about technology but also about the profound impact it has on society. By integrating his engineering background with his advocacy for respect, Christian Schenk embodies a holistic approach to innovation, ensuring that AI serves as a tool, fostering environments where both technology and humanity thrive together.

LinkedIn: https://www.linkedin.com/in/christian-schenk-respekt-training/

THE CENTRALITY OF DATA AND AI LITERACY: BUILDING CIVILIZATION'S NEXT COMMON LANGUAGE

By Amit Shivpuja
Senior Data & AI Executive, Author
Bentonville, Arkansas

An investment in knowledge pays the best interest.
—Benjamin Franklin

Literacy as the Foundation of Progress

Every era of human progress has been defined by a new form of literacy. The printing press democratized reading and writing, fueling the Enlightenment. The Industrial Revolution demanded technical literacy, enabling workers to operate machines and engineers to design them. The digital age required computer literacy, empowering billions to navigate the internet and software.

Today, as I argue in my book *The Data and AI Compass*, we stand at the threshold of a new epoch. Data and AI literacy is becoming part of the common language lexicon of our time. It is no longer a niche for technologists, engineers, etc. It is a civic, cultural, and professional necessity. Without it, individuals and organizations risk being overwhelmed by complexity, misinformation, and fear. With it, they unlock trust, innovation, and resilience.

The Literacy Gap: Why Fear Persists

In workshops, sessions, and discussions, I often ask leaders a simple question: "How confident are you in interpreting the data behind your most important decisions?" The silence and few responses that follow are telling.

Despite billions invested in analytics and AI, most organizations remain data illiterate at scale. Dashboards are misread. Correlations are mistaken for causation. AI models are deployed without understanding their assumptions or limitations.

This gap explains why fear dominates so many conversations about AI. Fear of job loss. Fear of bias. Fear of the "black box." These fears are not always irrational. They are the natural response to a lack of literacy. As I have written on my blog, *Data Compass* Substack, fear thrives in the absence of understanding.

Defining Data and AI Literacy

True literacy in this era has two intertwined dimensions.

1. **Data Literacy**—the ability to read, interpret, and question data. This includes understanding metrics, recognizing bias, and asking the right questions of analysts.

2. **AI Literacy**—the ability to understand how AI systems learn, predict, and act. This does not mean coding neural

networks, but it does mean grasping concepts like training data, model drift, and explainability.

On LinkedIn, I have argued that data literacy without AI literacy is incomplete. Data tells us what has happened. AI projects what could happen. Together, they form the compass that guides modern decision-making.

Historical Parallels: Literacy as a Civilizational Shift

History shows us that literacy is never just a skill. It is a societal shift.

- Reading literacy unlocked democracy, enabling citizens to engage with ideas and laws.
- Technical literacy fueled industrialization, creating new classes of workers and entrepreneurs.
- Computer literacy reshaped economies, birthing the knowledge economy.

Now, data and AI literacy will determine who thrives in the twenty-first century. Nations, companies, and individuals who master it will lead. Those who do not risk marginalization.

Case Study: The Workforce Divide

In my essay "Data Literacy vs. Business Literacy: Which Skill Drives Better Decisions?" I explored how organizations often silo these skills. Business leaders may excel at strategy but lack data fluency. Data scientists may master algorithms but struggle to connect insights to business outcomes.

The future belongs to those who bridge this divide. Imagine:

- A supply chain executive who not only reads demand forecasts but also understands how the AI model was trained to deliver the same.

- A marketing leader who can challenge whether personalization algorithms reflect customer values and are ethical.
- A healthcare administrator who can interpret predictive models without blindly trusting them.

These are not "nice-to-have" skills. They are survival skills.

The Three Levels of Literacy

I, thus, propose three levels of literacy:

3. **Foundational Literacy**—Everyone in an organization should understand basic concepts: what data is, how it is collected, what AI algorithms are being leveraged, and why bias matters.
4. **Functional Literacy**—Managers and professionals should be able to interpret dashboards, question assumptions, and integrate AI insights into decisions.
5. **Fluent Literacy**—Executives and specialists should be able to evaluate AI strategies, govern responsibly, and communicate implications to stakeholders.

This tiered approach ensures that literacy is not confined to a few experts but distributed across the enterprise.

Governance and Literacy: Two Sides of the Same Coin

In *The Data and AI Compass*, I emphasize that governance and literacy are inseparable. Governance provides the guardrails. Literacy provides the driver's skillset.

- Governance ensures AI models are explainable.
- Literacy ensures leaders can interpret those explanations.

- Governance enforces ethical standards.
- Literacy ensures employees understand why those standards matter.

Without governance, literacy risks being misapplied. Without literacy, governance risks being ignored. Together, they create trust.

Practical Pathways to Building Literacy

6. **Executive Sponsorship**—Literacy must start at the top. When leaders model data- and AI-informed decision-making, it cascades through the culture.

7. **Embedded Learning**—Literacy is not a one-time workshop. It must be embedded into workflows, tools, and daily decision-making.

8. **Cross-Functional Dialogue**—Organizations should encourage conversations where business leaders ask technical questions and data scientists explain in plain language.

9. **Governance as Education**—Governance frameworks should not only enforce compliance but also teach teams why standards matter.

10. **Storytelling with Data**—As I often share, numbers alone do not persuade. Literacy includes the ability to tell stories with data that resonate with human values.

Literacy in Practice: Examples Across Industries

- **Retail**—Store managers who understand demand forecasting models can adjust promotions with confidence.
- **Healthcare**—Clinicians who grasp predictive analytics can better interpret patient risk scores.

- **Finance**—Analysts who understand model drift can prevent flawed trading strategies.
- **Education**—Teachers who understand AI tutoring systems can guide students more effectively.

These examples show that literacy is not abstract. It is practical, immediate, and transformative.

Beyond Fear: Literacy as Empowerment

Fear of AI often stems from misunderstanding. In my Substack post "Human Oversight or AI Autonomy?" I argued that oversight is only possible when humans understand what they are overseeing. Literacy transforms fear into empowerment. It allows employees to see AI not as a black box but as a partner.

When literacy spreads, conversations shift:

- From "Can we trust this model?" to "How do we validate and monitor it?"
- From "AI will take my job" to "AI will augment my role."
- From "Data is overwhelming" to "Data is a compass."

The Global Stakes

Data and AI literacy is not just a corporate issue. It is a societal one. Democracies will need literate citizens to navigate misinformation and algorithmic influence. Economies will need literate workers to adapt to automation. Communities will need literate leaders to ensure AI serves human values.

Literacy is the foundation of trust at every level: personal, organizational, and societal.

Expanding Vision: Literacy as a Cultural Movement

Literacy must be seen not only as a skill but as a cultural movement. Just as reading literacy became a marker of citizenship, data and AI literacy must become a marker of participation in modern society.

This requires:

- **Policy Support**—Governments must integrate literacy into education systems.

- **Corporate Responsibility**—Companies must invest in literacy as part of workforce development.

- **Community Engagement**—Nonprofits and civic groups must promote literacy to combat misinformation.

Literacy as the Edge

The title of this book, *The AI Edge*, is fitting. But the true edge is not in algorithms or infrastructure. It is in people who are literate in data and AI. Literacy is the multiplier that turns technology into transformation.

As I emphasize in my writing, literacy is not just about skills. It is about mindset. It is about curiosity, critical thinking, and the courage to ask better questions. In a world where AI will shape every industry, literacy is not optional. It is existential.

About the Author

Amit Shivpuja is a senior data and AI executive, published author of *The Data and AI Compass*, and global thought leader. He currently serves as director of data governance for merchandising at Walmart US, where he architects frameworks that transform data chaos into clarity and trust. Through his *Data Compass* Substack and LinkedIn thought leadership, Amit explores the intersection of agentic AI, governance, and literacy. He holds an MBA from UC Irvine, a postgraduate degree in AI from UT Austin, and leadership certification from Stanford. Amit is passionate about enabling organizations to scale responsibly, blending technical mastery with human-centered storytelling.

Website: www.daicompass.com

Blog: datacompass.substack.com

LinkedIn: https://www.linkedin.com/in/amitshivpuja/

LEADING THROUGH INFLECTION: CULTURE, CURIOSITY, AND THE HUMAN SIDE OF AI TRANSFORMATION

By Shelly Swanback
Board Director and Transformation Executive
Arvada, Colorado

Culture does not change because we desire to change it. Culture changes when the organization is transformed; the culture reflects the realities of people working together every day.

—Frances Hesselbein

The Compliment That Became a Compass

I'll never forget an informal conversation with a boss at one of my most defining career milestones. He let me know that one of my greatest strengths was being a curious leader, always learning and pulling people into my learning with me.

I was a senior leader at Accenture, but for me it was never about being the smartest person in the room (we had plenty of those!); it was about asking the best questions and bringing the room along with me.

Shortly after, the leader who would become my new boss asked me to help start Accenture Digital with one caveat: We both needed to be clear that I didn't actually know anything about "digital."

He wasn't wrong. What I did know was how to learn fast and lead through change. We were building a new business inside Accenture at a time when "digital disruption" was still fuzzy and no one could really define it. So, we made curiosity a team sport. We wove learning into everyday routines and made it visible. Three of my favorite questions for any meeting quickly became:

- "What have we learned?"

- "What surprised you?"

- "What are we going to do next?"

Those simple questions turned uncertainty into momentum. They helped us navigate a noisy, ill-defined tech inflection by focusing on something timeless: how people learn, adapt, and grow together.

Today, AI is having its turn. With huge promise and plenty of uncertainty, the technology is more powerful, and the emotions are sharper, than in the digital transformation era. Leaders are told, "Move fast or you'll be left behind," while employees quietly wonder, "Will this replace me?" Yet underneath this inflection point, the human questions are strikingly familiar.

The good news is that if you led through earlier waves of digital transformation such as web, mobile, and cloud, you're more prepared than you might think. The same muscles you built then are the ones you need now. The leaders who will thrive in the AI era won't just have a sharp AI roadmap. They will build the strongest culture for learning, experimentation, and human-centered change.

There is no shortage of technical manuals on AI. What leaders at all levels of the organization need just as urgently is a playbook for the human and cultural operating system that makes AI transformation actually work.

In this chapter, I'll provide some practical tips on three parts of your cultural operating system (Cultural OS): how you talk about AI, how you develop learning mindsets and routines, and how you set guardrails.

Echoes from the Digital Era

The last cycle of digital transformation taught us that technology change without cultural change fails. That lesson is even more true in the AI era. To lead through this inflection, you must intentionally build a cultural operating system that normalizes change, elevates curiosity, and treats learning as core work, not a side project.

We can apply key lessons from the digital transformation wave to AI:

- Tech-first, people-second fails. Projects that focused solely on implementing platforms rather than upskilling employees and changing ways of working stalled or under-delivered.

- Story and language shape reality. Teams who could answer, "What does digital mean for us?" moved faster than those stuck in buzzwords.

- Learning speed beats planning perfection. Organizations that created fast feedback loops learned their way into success.

While much of AI feels new and undefined, the underlying leadership challenge is familiar: "How do you help humans move from fear and confusion to agency and participation?"

Most humans and organizations resist change. People worry about their status, skills, and job security. Despite all the transformation in the digital era, most organizations' processes, incentives, and metrics still reward stability and predictability, not experimentation and learning. They certainly do not account for more AI-powered systems and agents that are designed to learn and improve over time.

As a leader, you must address these emotional and cultural realities to move your AI strategy from PowerPoint decks and pilots to real changes in daily behavior. Think of technology and AI as the apps and culture as the operating system. If your Cultural OS is not built for change, your AI apps will not run reliably. In practice, this comes down to how you talk about AI, how you and your organization learn, and how you set guardrails.

Just like digital transformation, AI transformation is less a tech project and more a culture project with technical implications. To be successful, leaders do not need to be AI experts. But they do need to be deeply curious and willing to make visible, day-to-day changes to build a Cultural OS for AI transformation.

Let's dive into some practical advice on what to do.

Cultural OS Component 1: Messaging and Communications

It is critical to talk honestly about AI, including both the promise and the potential downsides. What you say about AI, and how often people hear it, matters a lot. Most leaders are taking a grounded, optimistic approach as they shape their company-specific message. Laying out a vision for what the company aspires to be with AI is important. Just as important are concrete messages about what employees can expect, such as:

"You are going to see AI show up in more of our tools and processes. Our goal is not to replace people, but to remove repetitive, low-value tasks, so we can focus on more meaningful work. That means we will be experimenting together, and we will be transparent about what we are learning and where we are uncertain. We are asking you to lean into AI and be part of the learning and transformation."

When things are changing and leaders go quiet, people fill in the blanks themselves, often in anxious ways. It is surprising how often messaging has to be repeated before it really lands, especially when a topic is new, ambiguous, and emotionally charged like AI.

I remember a company-wide leadership meeting about two years into building Accenture Digital. I dreaded getting on stage and repeating what felt like the same story with a few new client examples. Thankfully, I did not change course. At that meeting I realized the organization was just starting to understand what digital disruption meant for their part of the business.

Here are some practical steps you can take for clear, frequent, and impactful communication:

- **Anchor everything to a clear, repeated theme.** Develop a simple, company-specific explanation of how AI will be good for your customers, your people, and your communities.

- **Use multiple formal and informal channels.** Share what is happening with AI at all levels. Regularly feature AI successes and learnings in town halls, and frame them as part of who the organization is becoming, not one-off projects. Use short, simple messages in existing channels (Slack or Teams, intranet, email) to reinforce that AI is an ongoing part of the business. Consider dedicated AI learning channels for people who want more depth.

- **Make sure communication channels are two-way.** Give employees opportunities to ask questions and get honest, straightforward responses.

- **Create a simple learning series.** One of the most powerful things I did leading a global transformation was to send out a Question of the Month to get people thinking about a challenge or learning something new. Questions were simple so employees at all levels could respond. For example:

- o "What do you wish you could automate or take away completely to make your job easier?"
- o "What is your favorite personal use case for ChatGPT or other AI tools, and could any part of it apply at work?"

For AI messaging to stick, it has to be translated, not just broadcast. The CEO sets the overall story, but leaders throughout the organization make it real by putting it in their team's language and context. For example, each leader can create two simple team narratives:

- **"Here's what AI means for our team."** Translate the enterprise theme into your team's language. Be specific about impacts to customers, processes, and KPIs.

- **"Here's what will and won't change for you in the next six to twelve months."** Be concrete about what people can expect, where you're experimenting, and what's off the table right now.

If you lead a function inside a larger organization, explicitly connect your team narrative back to the company story: "You've seen the CEO's note about AI. Here's what that means for our team this quarter ..."

One of the best ways to assess whether your messaging and channels are working is to ask employees to answer, in one sentence, "What does AI mean for our organization? What does it mean for your team?" If they struggle to respond or the answers are all over the map, treat that as useful feedback. It is a signal to further sharpen and repeat your messaging.

Cultural OS Component 2: Learning Mindsets and Routines

If you want a curious, always-learning culture, you have to design learning into your company rhythms. Formal learning programs will play an important role, but AI is changing quickly, and no one can stay current with a single training or an annual leadership offsite. You need simple, repeatable routines that make learning and experimentation a normal part of daily work. Learning has to be understood as part of the job, not a side project or an annual checklist item.

Start with basic AI literacy for everyone, so the organization has a shared understanding and vocabulary. Fortunately, many AI 101 programs are readily available. As more people use AI in their personal lives, you can tap those examples and metaphors to help your employees explore work use cases.

Changing mindsets takes time, repetition, and alignment from the top to the bottom of the organization. Here are some ways to rewire your company routines (or your team's routines) to make learning visible.

Make Status Reports Learning Artifacts

It is powerful and empowering for your team to add two standing questions to written and verbal status updates:

- "What did we learn this week?"
- "What will we do differently because of that learning?"

Over time, this normalizes change and experimentation. It shifts the focus from only "Are we on track?" to also "Are we getting smarter?" This will be particularly important as we deploy more AI agents that are designed to learn and improve over time.

Design Meetings with a Learning Segment

Reserve the first or last 10 minutes of recurring meetings for a quick AI demo from a team member or a short "experiment highlight" that covers what was tried, what happened, and what was learned. Bringing employees as "special guests" to meetings outside their day-to-day can also be powerful. One of my favorite examples is a meeting where

I brought a finance manager to a marketing session about revising our client messaging. The finance manager did not question spend. Instead, they suggested ways to ensure our messaging resonated with CFOs as well as operational executives.

Show Up and Engage in Informal Learning and Experimentation

You might already have recurring AI office hours where people bring a problem and get help using AI tools to think it through, ideally with rotating hosts such as product managers, data scientists, and "AI power users." Learning spreads quickly when senior leaders show up to these sessions and engage in the problem solving. There is no better way to walk the talk than to work side by side with teams in informal learning and experimentation.

Evolve Measurements to Celebrate Learning, Not Just Outcomes

We tend to get more of what we measure. The good news is that celebrating and rewarding learning behaviors is straightforward. Examples include adding "learning contributions" to performance conversations (who is sharing best practices, mentoring, or creating internal guides), featuring "failed" AI experiments and their lessons in town halls, and tracking the number of AI experiments attempted, not just the successful ones. One of my favorite ways to reinforce learning is to use company forums to casually thank an employee for teaching me something or shedding light on a new insight.

In the end, if you want an always-learning culture, you must make learning part of the core work and design it into the daily calendar, not just the annual priorities or the strategy deck.

Cultural OS Component 3: Practical Guardrails

As regulations and ethical boundaries for AI evolve, your employees will keep seeing stories about AI missteps. It is critical to set up a clear AI lane with guardrails—simple principles, visible rules, and clear decision rights—so people know what is allowed and where to ask for help.

Employees are far more likely to experiment with AI when they trust two things: that the company cares about using it responsibly and that they will not get in trouble for trying something in good faith. Clear guardrails create that trust. They answer questions such as: What data is off limits? When does a human always make the final call? Who do I ask if I am unsure?

Many companies fall into the trap of forming a massive committee that spends a year writing policies while the rest of the organization waits or experiments in the shadows.

You are better off starting small and moving quickly:

- Establish a small cross-functional group (legal, IT, HR, and business leaders) that sets initial guardrails, reviews edge cases and incidents, and iterates policies based on real usage. This group should model the use of the guardrails in their own work and be available for quick-turn questions.

- Empower local teams to run small, low-risk experiments within the established guardrails, and encourage them to propose updates based on their experiences.

The cross-functional group should establish a few core principles to help navigate rapidly evolving regulations and ethical considerations. These might address:

- Protecting confidential and personally identifiable information

- Being transparent with customers when AI is used in interactions or content

- Complying with industry-specific regulations

- Ensuring humans remain accountable for high-stakes decisions

- Committing to use AI in ways that benefit customers and society, not just short-term efficiency

Again, employees respond best when they can see what the rules look like in practice: which AI tools are approved for which uses and when human review of AI-generated output is required. A simple "AI Dos and Don'ts" guide can go a long way. For example:

- Do: Use AI to draft internal emails, summarize documents, and brainstorm options.

  Do not: Use AI to make final decisions about hiring, firing, or performance evaluations.

- Do: Use AI to generate first drafts for customer communications.

  Do not: Send AI-generated content to customers without human review.

For people leaders, this means knowing the core guardrails and translating them into "on the ground" examples: what's in-bounds and out-of-bounds for your data, your processes, and your customer interactions. When someone brings you an edge case, resist saying, "Legal won't like that." Instead say, "Let's walk through the principles and, if needed, ask the AI cross-functional governance group together."

Good AI governance should feel like a well-marked lane with guardrails, not a maze of red lights. When people see leaders following the rules themselves, AI feels less like a risk being pushed on them and more like a tool they can trust.

Whether you're the CEO or a leader at any level, your messaging, learning routines, and guardrails form the Cultural OS your people will live in as AI shows up in their work. You don't have to rebuild everything at once. Start small and build from there.

Call to Action

AI will not just change workflows; it will change what it feels like to work in your organization. It can either shrink people, making them feel replaceable and powerless, or it can expand them, giving them new tools and new ways to contribute. As a leader, you decide which direction it goes. The choices you make about how you communicate, how you and the organization learn, and the guardrails you set will shape your employees' experience every day. And you can take simple action now:

- Change one routine. This week, add a few questions to your next team meeting: "What is something new we tried?" "What did we learn?" and "What would we like to try next?" Capture the answers and commit to one small experiment.

- Convey one authentic message. Send a short video or note to your team about AI. Share what you're excited about, and also what you wonder or worry about, and how you plan to address that. If you're a CEO, do this for the whole organization. If you're a people leader, do it for your team or function and tie it explicitly back to the company's AI message.

You don't have to have all the answers about AI. Your job is to create a culture that can live inside the questions and learn fast enough to turn uncertainty into possibility. That starts with a new conversation, not a new tool, and you can start that conversation this week.

About the Author

Shelly Swanback is a board director and transformation executive who has spent her career helping global organizations navigate digital disruption and large-scale cultural change. As the former Accenture Digital Group Operating Officer, she helped build and scale the firm's flagship digital business and later led organizations like Western Union and TTEC through major transformation initiatives. She now serves on the board of directors of WTW and advises leaders on AI, culture, and organizational change, while finding fresh perspectives traveling, hiking, and skiing in the Colorado mountains.

Email: shelly.swanback@outlook.com

LinkedIn: https://www.linkedin.com/in/shellyswanback/

CHAPTER 26

VIBE CODING WITH AI: WHEN CREATIVITY LEADS, LEARNING FOLLOWS.

By Sakina Syed
Microsoft IT Professional, AI Engineer and Consultant
Toronto, Canada

Coding is today's language of creativity. All our children deserve a chance to become creators instead of consumers of computer science.

—Maria Klawe

Vibe Coding with AI: Your Creative Edge

Ever thought to yourself, "I have this great app idea, if only I knew how to code"? You're not alone. For many, the gap between imagination and execution feels huge. That's where vibe coding comes in: a fresh, creative approach to programming that makes coding feel less like a chore and more like an art form. But what exactly is vibe coding? Why can it give you an edge in today's AI-driven world? What do you need to get started? And most importantly, how do you truly benefit from it? In this chapter, we'll explore how combining vibe coding with AI can unlock your creativity, accelerate learning, and help you turn ideas into interactive experiences without drowning in syntax or complexity.

If you're wondering, "What is vibe coding?" let me explain. Vibe coding is a creative, playful approach to programming that shifts the focus from rigid syntax and technical perfection to aesthetics, interactivity, and fun. Instead of starting with dry exercises or complex algorithms, vibe coding encourages developers, especially beginners, to dive into projects that feel exciting and visually rewarding. Imagine coding as an art form where you can create colorful animations, interactive games, or even music-driven visuals. This approach transforms programming from a purely logical task into an expressive experience, making it more accessible and enjoyable.

Why Vibe Coding Matters

The Intimidation Factor: Ever tried coding and felt completely stuck? For many beginners, programming can feel like learning a foreign language filled with cryptic symbols and rigid rules. Syntax errors, logic hurdles, and steep learning curves often create frustration long before creativity has a chance to shine. Traditional methods—memorizing commands and spending hours debugging—can drain motivation and discourage exploration. The result? Brilliant ideas stay locked in imagination because the process feels overwhelming.

Flipping the Script: Vibe coding changes that narrative. Instead of starting with abstract theory or endless tutorials, it begins with creativity. You dive straight into building something visually engaging or interactive, like a colorful animation or a simple game, without worrying about perfection. Once the excitement kicks in, learning the logic behind the scenes becomes natural. This reversal of the traditional approach makes coding feel less like a technical chore and more like an expressive art form.

How AI Amplifies the Experience and the Edge It Gives You

Artificial intelligence takes vibe coding to the next level by acting as a creative partner. Tools like GitHub Copilot, ChatGPT, and Replit Ghostwriter can generate starter code, suggest improvements, and explain concepts in plain language. This means beginners can spend less time wrestling with syntax and more time exploring ideas. AI enables rapid prototyping, turning "What if?" moments into working projects within minutes. By reducing complexity and offering real-time guidance, AI makes vibe coding not only fun but also deeply educational.

In a world where technology evolves at lightning speed, vibe coding with AI offers a unique edge. It blends creativity with technical skill, helping developers learn faster while building projects that stand out. Whether you're dreaming of an app, a game, or an interactive art piece, vibe coding empowers you to bring those ideas to life without waiting years to master every detail. It's not just about learning to code. It's about learning to create, innovate, and thrive in an AI-driven future.

Immediate Feedback Sparks Motivation

One of the biggest advantages of vibe coding is instant gratification. When you write a few lines of code and see a shape move, a color change, or music react on your screen, the feedback loop is immediate. This visual or interactive response keeps learners motivated and curious. Instead of staring at error messages, beginners experience progress in real time, which builds confidence and encourages deeper exploration.

Vibe coding thrives on personalization. You're not just following a rigid exercise. You're creating something that reflects your style and imagination. Want a neon color palette? Add it. Prefer smooth animations? Tweak the timing. This freedom to experiment fosters creativity and problem-solving skills in a low-pressure environment. Mistakes become opportunities to learn, not roadblocks to success.

Ultimately, vibe coding matters because it transforms learning into an enjoyable journey. By focusing on creativity first, beginners develop core programming skills—loops, conditionals, functions— without feeling overwhelmed. They learn by doing, experimenting, and asking questions, which is far more effective than memorizing syntax. Combined with AI assistance, vibe coding becomes a powerful gateway to both technical competence and creative confidence.

The Role of AI in Vibe Coding

AI as Your Creative Partner: Artificial intelligence transforms vibe coding from a solo activity into a collaborative experience. Instead of struggling through syntax or searching for obscure documentation, AI acts as a creative partner that helps you bring ideas to life quickly. Whether you want to design a colorful animation, build a simple game, or create music-driven visuals, AI can suggest code snippets that fit your vision. This partnership allows beginners to focus on creativity while AI handles the technical heavy lifting.

One of the most powerful aspects of AI in vibe coding is its ability to explain complex programming concepts in simple, human-friendly terms. Instead of reading dense documentation, you can ask AI, "What does this loop do?" or "Why do I need a function here?" and get clear, concise answers. This makes learning feel conversational rather than intimidating, turning coding into an approachable skill rather than a technical barrier.

Debugging Without Killing the Flow: Nothing disrupts creativity faster than hitting an error wall. Traditionally, debugging can consume hours and drain enthusiasm. AI changes that by offering quick fixes and explanations without breaking your creative momentum. If your animation freezes or your game logic fails, AI can pinpoint the issue and suggest solutions instantly. This keeps the vibe alive, so you can stay focused on building, experimenting, and enjoying the process.

Tools That Make It Possible: Examples of Vibe Coding Tools

Several AI-powered tools make vibe coding accessible and fun. GitHub Copilot provides real-time code suggestions as you type, helping you write functional code faster. ChatGPT excels at brainstorming ideas, generating snippets, and explaining concepts in plain language. For collaborative coding, Replit Ghostwriter allows teams to work together while leveraging AI assistance. These tools don't just speed up development. They make coding feel like a creative dialogue rather than a solitary struggle.

SteerCode is another powerful application that enables users to generate complete applications simply by providing natural language prompts. Beyond code generation, it also allows users to design and configure entire development environments, streamlining the process from concept to deployment.

While AI can generate code instantly, the real value lies in understanding what that code does and learning by asking why. Beginners should resist the temptation to copy-paste blindly and instead ask AI questions like, "Why did you use this loop?" or "What happens if I change this variable?" This active engagement turns AI from a shortcut into a mentor, helping learners build a strong foundation while still enjoying the creative freedom vibe coding offers.

AI doesn't just make coding easier. It redefines what coding can be. By blending automation with creativity, vibe coding powered by AI opens doors for anyone with an idea, regardless of technical background. It's a glimpse into the future where coding is less about memorizing syntax and more about expressing imagination. For beginners, this means learning faster, creating more, and enjoying every step of the journey.

Getting Started with Vibe Coding + AI

Begin with the right platform. The first step in vibe coding is choosing a creative platform that makes coding fun and visually rewarding. For beginners, popular options include p5.js, a JavaScript library for interactive graphics; Processing, which is great for visual art and animations; and Python Turtle, a beginner-friendly tool for drawing shapes and patterns. These platforms are designed for creativity, offering simple commands that produce immediate visual results, which is perfect for anyone who wants to see their ideas come to life quickly.

As you grow more confident, you can explore more advanced environments that expand your creative possibilities. Tools like Unity and Unreal Engine allow you to build immersive 2D and 3D experiences, while Blender's scripting capabilities let you combine coding with professional-grade animation and modeling. For web-based interactivity, Three.js enables stunning 3D graphics in the browser, and TouchDesigner offers a node-based approach for generative visuals and interactive installations.

AI-powered platforms also play a huge role in vibe coding. Lovable is a fantastic option for beginners who want to build apps without getting bogged down in complexity. It uses AI to generate functional, visually appealing applications with minimal effort.

Similarly, Claude, an AI assistant by Anthropic, excels at explaining code, brainstorming creative ideas, and helping debug without breaking your flow. These tools, combined with traditional coding platforms, open doors to complex projects like virtual reality experiences, interactive art installations, and even AI-driven games, giving you the freedom to scale your creativity from simple sketches to full-blown digital worlds.

As mentioned above, SteerCode is another fantastic application to build apps from scratch with basic prompts. What makes it stand out is its ability to turn simple natural language instructions into fully functional applications, while also giving you control over the development environment. This means you're not just generating code. You're shaping the entire creative ecosystem for your project. For vibe coders, this opens up endless possibilities: from rapid prototyping to scaling ideas into polished, production-ready experiences without sacrificing creativity.

Let AI scaffold your project. Once you've picked your platform, it's time to bring AI into the mix. Instead of starting from scratch, use AI to generate a basic structure for your project. For example, you might ask: "Create a colorful animation with bouncing shapes in p5.js" or "Write Python Turtle code for a spiral pattern." AI will provide a working foundation, saving you time and reducing frustration. This approach allows you to focus on the creative aspects rather than getting stuck on syntax.

For the examples above, here are some sample prompts you can start off with:

- "Create a bouncing ball animation in p5.js."
- "Draw a colorful spiral using Python Turtle."
- "Generate a simple interactive game with keyboard controls."

The real magic of vibe coding happens when you start tweaking and personalizing your project. Change the colors, adjust the speed of animations, or add interactive elements like mouse clicks or keyboard controls. These small modifications teach you how variables, loops, and conditions work without feeling like a rigid lesson. AI can help here too: Ask for suggestions like "How do I make the shapes change color when clicked?" and watch your project evolve.

Ask AI to explain the logic. Don't just copy and paste. Use AI as a mentor. After generating code, ask questions like "Why did you use a loop here?" or "What does this function do?" AI can break down concepts in plain language, making it easier to understand programming fundamentals. This active learning approach ensures you're not just building cool projects. You're also gaining the skills to create independently in the future.

Experiment without fear. One of the biggest advantages of vibe coding with AI is the freedom to experiment. If something breaks, AI can help debug quickly, so you never lose your creative momentum. Try bold ideas: add sound effects, create random patterns, or make your animation respond to user input. Every experiment teaches you something new, and with AI by your side, mistakes become stepping stones rather than roadblocks. And while creativity is the focus, don't forget security. Always review AI-generated code for potential risks, such as hardcoded credentials or missing input validation, to keep your projects safe.

Build confidence through play. Starting with creativity instead of complexity builds confidence. When you see your code produce something fun and interactive, you feel empowered to keep going. Over time, you'll naturally absorb programming concepts like loops, conditionals, and functions because you've seen them in action. This playful approach turns coding from a technical challenge into an enjoyable journey.

Do remember to post your vibe coding experiments online on platforms like GitHub or community forums because feedback and collaboration can spark new and improved ideas.

Your First Step Toward Innovation

Getting started with vibe coding and AI isn't just about learning to code. It's about unlocking your ability to innovate. By combining creativity with AI-powered assistance, you can turn ideas into reality faster than ever before.

Whether you dream of designing interactive art, building games, or creating apps, vibe coding gives you the foundation to thrive in an AI-driven world. So pick your platform, fire up your imagination, and let AI amplify your creative edge! Start with one creative prompt today, your first step toward building something extraordinary.

Fun Project Ideas

Colorful Mouse Tracker: Start simple with a Colorful Mouse Tracker, where circles follow your mouse across a rainbow background. This project is perfect for beginners because it introduces concepts like event handling and dynamic positioning in a visually rewarding way. You'll learn how to track mouse coordinates and update shapes in real time, creating an interactive experience that feels magical. With AI's help, you can scaffold the code quickly and then tweak colors, sizes, and patterns to make it uniquely yours.

AI-Assisted Music Visualizer: If you love music, an AI-Assisted Music Visualizer is a fantastic way to combine sound and visuals. Using libraries like Tone.js, you can create shapes that pulse and react to beats, turning audio into art. AI can help you set up the audio analysis and animation logic, while you focus on customizing the visuals: choosing colors, shapes, and effects that match your vibe. This project teaches timing, arrays, and event-driven programming in a fun, immersive way.

A visualizer area can include:

- Neon-colored concentric rings pulsing to bass frequencies.
- Rainbow bars at the bottom reacting to mid/high frequencies.

- Floating particles triggered by beat detection (spectral flux peaks).

Snowfall Animation: Bringing Winter Magic to Your Screen: For seasonal flair, a Snowfall Animation is the perfect project to capture cozy winter vibes and showcase the beauty of creative coding. Picture delicate snowflakes drifting gracefully across a serene background, each flake varying in size, speed, and opacity for a realistic effect. This project introduces essential programming concepts like loops for generating multiple snowflakes, randomness for creating natural variation, and layering for depth and perspective. With AI as your coding partner, you can scaffold the initial design in seconds. Simply prompt, "Create a snowfall animation in p5.js with random snowflake sizes and speeds."

From there, the possibilities for customization are endless. Add wind effects to make flakes sway, incorporate interactive elements where snowflakes melt on click, or even sync the snowfall to background music for an immersive experience. Feel free to add different snowflake shapes later. These enhancements not only make the project visually stunning but also teach advanced concepts like event handling and physics simulation in a fun, low-pressure environment. By blending creativity with AI-powered assistance, you'll transform a simple animation into a dynamic winter wonderland that reflects your unique style.

Mini Game: Catch the Ball: Ready for a challenge? Build a Mini Game: Catch the Ball, where players move a paddle to catch falling objects. This project teaches collision detection, scoring systems, and user input handling, all essential skills for game development. AI can help you structure the game logic, then you can add your own twist, like power-ups, changing difficulty levels, or vibrant themes. It's a fun way to learn while creating something playable. With AI's help, you can scaffold the basic structure in seconds. Just prompt: "Create a p5.js game where a ball bounces and the player controls a paddle."

Mood-Based Backgrounds: Finally, explore Mood-Based Backgrounds, where AI picks colors based on your text input. Type "calm," and watch soothing blues appear; type "energetic," and see bold reds and yellows take over. This project blends natural language processing with creative coding, showing how AI can interpret emotions and translate them into visuals. It's a perfect example of how vibe coding and AI work together to create personalized, interactive experiences.

Best Practices

Think of AI as your mentor. To get the most out of vibe coding with AI, treat AI as a mentor, not a shortcut. Instead of blindly copy-pasting code, ask questions and understand the logic behind each suggestion. Embrace experimentation and don't be afraid to break things. Failure is part of the creative process and often leads to the best learning moments. Finally, share your projects with others. Collaboration not only makes vibe coding more fun but also exposes you to new ideas and techniques that can spark even greater creativity.

Keep security in mind. While vibe coding emphasizes creativity and experimentation, security should never be an afterthought. When using AI tools to generate code, always review the output for potential vulnerabilities such as hardcoded credentials, insecure API calls, or missing input validation. Avoid blindly trusting generated code. Test it thoroughly and apply best practices like sanitizing user input, using secure libraries, and following the principle of least privilege. Remember, creative projects can still be targets for exploitation, so building with security in mind ensures your work remains safe and reliable.

Vibe coding with AI is more than just learning syntax. It's about unlocking creativity, confidence, and a sense of play in programming. By starting with visually engaging projects and letting AI guide you through the technical details, you transform coding from a rigid task into an expressive, enjoyable experience. This approach prepares developers for a future where humans and AI co-create solutions, blending imagination with intelligent automation. So start small, stay curious, and let AI amplify your ideas and give you an AI edge! The next big innovation might just begin with a few lines of code and a spark of creativity. Even something as simple as typing a basic prompt like "Create a bouncing ball animation" can set your creative journey in motion.

About the Author

Sakina Syed is a dynamic senior data and AI consultant, AI engineer, and enthusiast with a stellar track record in the computer software industry, including notable tenures at tech giants like Microsoft. As a Microsoft AI engineer and Azure-certified professional, she excels in AI deployment, sales, management, teamwork, and leadership.

With a Bachelor of Science in Neuroscience and Mental Health Studies from the University of Toronto, Sakina has also enriched her expertise with courses in artificial intelligence, business management, and IT. Her passion for writing and traveling adds a unique dimension to her professional persona.

Sakina's enthusiasm for business and the application of business psychology to understand consumer needs is matched by her extensive seven-year experience in customer service. She is adept at interpreting consumer feedback and statistical data, enabling her to identify market requirements with precision. Her strong passion for team management and leadership drives her dedication to delivering exceptional customer experiences and fostering business success.

Email: sakina@youraiconsultant.ca

Website: https://youraiconsultant.ca/

LinkedIn: https://www.linkedin.com/in/sakina-syed/

CHAPTER 27

DE-RISKING AI IN A GEOPOLITICAL RACE FOR TECHNOLOGY DOMINATION

By Leszek Tasiemski
AI Security Expert
Poznań, Poland

From Promise to Precarity

We are in an AI arms race. It matters on business, government, and personal levels. This miracle technology, promising a human-like companion, brilliant programming, and vehicle autonomy, also happens to be the most potent technology for disinformation, cyberattacks, inventing new biological weapons, mass-surveillance, or fully autonomous lethal weapons. Nation states and Big Tech know that this is the race to win. Whoever gains AI dominance will gain leverage to dominate also in other areas: the economy and research, but also the military and geopolitics.

If you are reading this chapter, it means you are at least mildly interested in the darker side of AI. Let me offer you a global perspective on the AI-related risks with a special focus on the European Union, of which I'm a citizen. Global risk management in such a quickly moving landscape as AI is tricky, yet some clear patterns and de-risking strategies already emerge.

The Expanding Risk Landscape

AI comes with a rapidly expanding risk landscape. Those risks are related to misuse and attack techniques against AI-enabled systems or vulnerabilities in AI systems, leading to issues like misalignment (so, when AI may pursue its own agenda).

Misuse is what worries me most. Powerful models and agents can conduct an autonomous exploitation of computer systems to gain access, steal data, or sabotage or conduct fraudulent financial operations. Influence campaigns are already done on an unprecedented scale, with such quality that it's becoming an impossible task to distinguish between fact and fiction, especially when hybrid methods are used that blend AI-generated content with some in-real-world actions done by small groups of people.

Increasingly, AI disinformation produced in great quantities is not intended for human consumption. It's intended to be ingested by other AI models to poison them. Recent research has proven that it's surprisingly easy to skew a large language model.

AI is also pivotal in military (and potentially, terrorist) activity. It's not only about military decision support systems (DSS) that assist in mission planning, target identification, and similar tasks. Fully autonomous lethal weapons are already a fact in countries that value military power over human rights and safety. We need to be aware of the risk of AI guiding research on novel biological or chemical weapons.

Remember: What is a risk in one context would become a feature in the other. Most assessments, group the core AI risks into four categories:

- Persuasion: influencing beliefs and actions of humans
- CBRN (chemical, biological, radiological, nuclear) weapon creation
- Cybersecurity: meaning the ability to autonomously detect weaknesses and conduct attacks
- Model autonomy: resistant or unsupervised behavior, often referred to as misalignment

While all the above characteristics of an AI model are undesirable for the public and nations, this is exactly the kind of feature that could bring a decisive advantage in a conflict or a war. So, dominance in the AI field is not just about business and technological progress. With that in mind, it's easy to understand the rhetoric and behavior of the players on the geopolitical scene, notably the USA, strongly resembling the Space Race in the 1960s and 70s or even the Manhattan Project. Not all countries and blocs are openly discussing their own ambitions, but everyone understands very well that this is the one to win.

The Geopolitics of AI Dominance

What does an AI dominance really mean in practical terms? To have a chance of succeeding, four ingredients are crucial. You need access to data, specialized hardware to process it, electricity to power the hardware, and, finally, human experts to convert the ingredients into a superior model or a swarm of models.

In terms of economic importance, there is no exaggeration in the saying that "data is the new oil." Without sufficiently good data, in terms of both quality and quantity, the success in the AI field is impossible. What kind of data is needed depends on the purpose of the model, but we can safely assume that the two biggest obstacles are privacy laws and copyrights.

While the first generation of LLMs ingested all the data available on the internet, we can see an increasing number of regulations and restrictions in usage of data to train the models. We (at least in Europe) are given an option to opt out from using our content to train or tune AI models. Europe is leading the way, regulating and requiring vendors to document the sources of data used in training.

For obvious reasons, not having to care about privacy or copyright is an asset from the AI race perspective. Aggressive data collection and less restrictive privacy and copyright norms give Chinese AI entities a competitive edge. Accusations of using a competitor's model to train your own AI, known as model stealing, or using restricted content for training are quite common. For instance, around January 2025, OpenAI has alleged that the Chinese AI company DeepSeek may have used a technique called "distillation" to train its own models using outputs from OpenAI's systems, which would violate OpenAI's terms of service. The situation has prompted broader discussion about data use practices in the AI industry, as OpenAI itself has faced criticism regarding its approach to data collection for training its own models. Apart from that, techniques such as social media scraping through "AI personas" are also observed.

Assuming you have the data, you need hardware to train an AI model. Typically, vast data centers are needed, with networking, storage, and cooling, but the scarcest assets are GPUs, graphical processing units. Such hardware was initially created to boost performance in graphics-related calculations, compared to CPU, but over time, it became clear that GPUs can very efficiently run calculations needed by AI. The US-based company NVIDIA practically monopolized the market of high-performance AI chips. Other big players on the processor market, like AMD or Intel, also have a limited GPU offer, but as of now, NVIDIA still dominates.

Interestingly, the government of the United States, being aware of the competitive advantage, began capping the sales of high-performance GPU chips to other countries and regions, notably, impacting China. In mid- to long-term, this may play into China's favor. Having no other choice, China is developing the Ascend series of GPUs, produced by Huawei. Currently, the Ascend chips still are roughly one generation behind equivalent chips from NVIDIA.

From a geopolitical perspective, two more factors are important. Even though NVIDIA is a US company, the vast majority of the chips are produced by TSMC, Taiwan Semiconductor Manufacturing Company. It matters a lot, especially considering political tensions between China and Taiwan, as well as the possibility that the US military would defend Taiwan in case of a potential forceful takeover attempt.

Europe doesn't play an important role in the GPU scene with one exception. There is no notable European GPU industry, but there is ASML. It's a Dutch company that is a leading supplier of advanced photolithography systems. It provides chipmakers with the hardware, software, and services needed to create the tiny structures on silicon that form processors, like GPU. There is an interesting dependency here, because even the biggest manufacturers, based in the US, producing in Taiwan, still critically depend on a Dutch company in their supply chain. ASML, under Dutch export-control rules, can't sell the latest generation of systems, like extreme ultraviolet (EUV) scanners, to China. Those systems are currently off-limits for Chinese chip factories, which makes catching up with NVIDIA even harder.

India took an interesting approach to AI and hardware. One of its startups is specializing in adapting AI models to efficiently run on standard (and much less expensive) x86 architecture processors.

There is one more interesting angle when it comes to hardware and the AI race: location. AI datacenters will become critical assets and among primary targets in a potential armed conflict or terror/sabotage act. This may be one of the motivations behind China's Three-Body Computing Constellation project that aims to build a distributed AI data center in orbit. The first 12 out of 2,800 planned satellites of the constellation were launched in May 2025. It is significantly harder for any adversary to take such a data center down compared to anything based on the ground.

The third critical asset to think of AI dominance is energy. AI data centers consume vast amounts of electricity, both for training and inference (using) of the model. Modern hyperscale AI clusters require hundreds of megawatts, while gigawatt-scale ones are being planned.

Such demand causes energy price hikes in the regions where such data centers are built. Interestingly, there are also projects to restart some of the decommissioned nuclear power plants in the USA. Those include Three Mile Island Unit 1 in Pennsylvania to supply power for Microsoft data center operations and Duane Arnold Energy Center in Iowa to power Google's cloud and AI infrastructure.

The final ingredient is human talent. Experts are needed to use the chips and energy to turn the data into an AI technology, be it model, agent, or a combination of both. In the commercial sector, we can observe an aggressive fight for the top talent, with NBA-worthy multi-million-dollar transfers. Notably there were recent transfers between Microsoft and Google DeepMind and Meta's "superintelligence" hiring spree. Those transfers make news because commercial entities are involved. Such acquisition of experts is happening around the world, also at government and military entities.

Countries and the European Union have their initiatives and strategies to secure a dominant position, if not overall, then at least to settle into an AI niche. In March 2024, India announced its AI Mission initiative. January 2025 marked the US Stargate initiative with a joint venture committed to invest up to US $500 billion in AI infrastructure over the next four to five years. In February of the same year, the European Union held an AI summit in Paris, followed by the EU AI Continent Plan announced in April 2025, with up to a €200 billion investment perspective that will include the creation of AI factories. Investment matters, considering that the cost of training an LLM can be over US $100 million.

The European Way

In the global AI race, the European Union is seeking its own position and niche. The position is defined through various initiatives, like the aforementioned EU AI Continent plan or the AI Act. While the tricky pursuit for balance between regulation and boosting innovation continues, the fundamental principles of the EU approach are very clear. Safety, fundamental rights, transparency, and privacy stand out in all the initiatives. EU objectives include sovereignty and reduced dependence on US and Chinese AI giants in terms of foundational models, cloud infrastructure, chips, platforms, and datasets.

Critics point out that the "regulate first" approach further slows the European progress down; however, it also restricts the Big Tech players from gaining a dominant position by abusing data of their users. Such an approach opens the market for AI vendors who design and build their datasets and models with safety and privacy principles in mind. To lower the bar for startups, the EU's AI Continent strategy explicitly includes organizational structures and support mechanisms designed to help SMEs (small and medium-sized companies) both comply with new rules (like the AI Act) and stay competitive in an AI-driven economy.

The EU AI Act uses a four-tier, risk-based classification system. Every AI system placed on the EU market must fall into one of these categories. Notably, the classification starts with unacceptable risk applications (tier 1), defining which AI systems are prohibited in the EU. Such prohibited systems include social scoring done by governments, manipulative or subliminal techniques that affect behavior, real-time biometric identification in public areas, and systems exploiting vulnerabilities of children. Such systems cannot be used or sold in the EU.

Then, the classification continues with high-risk systems (tier 2), like the medical applications of AI, to limited risk and no risk systems (tiers 3 and 4). What is crucial, the requirements in terms of transparency, logging, accountability, documentation, cybersecurity, or monitoring progress form almost no burden in the no risk category to significant obligations for the high risk systems. This way, the bureaucratic burden is scaled up along with the damage potential of the specific AI product.

Europe's AI strategy focuses on safety and de-risking AI, while building independence from today's dominant market players in terms of foundational models, data, and infrastructure. With adequate levels of regulation, it is possible to create a robust market for safe AI models, with transparency regarding training data and methods.

Threat Actors Don't Care About Laws

While a safety and privacy first approach, like that presented by the EU, is very effective to prevent misuse of data by legally operating businesses and governments, please remember that threat actors simply don't care.

Threat actors range from financially motivated criminals, to hacktivists, to state-led hacking groups, referred to as APTs (advanced persistent threat operators). They all operate outside the law; hence no regulation would restrict them. To some extent, the usage of AI tools by such actors is currently limited by the safety mechanisms built into the models. For instance, none of the leading AI tools would easily assist us in building malware or preparing a disinformation campaign. Even if a hacker bypasses such mechanisms, which happens routinely, cloud-based AI systems are being monitored for malicious usage and such operators are exposed. This recently happened with Anthropic, revealing that its AI product was abused by a China-linked APT group to breach several third-party systems.

In the near future, threat actors will proceed to mostly use private models, either with removed security mechanisms (jailbroken) or custom trained to perform malicious operations. When this happens, the technology companies and law enforcement agencies will lose visibility into such operations, which will significantly complicate detection and takedown. Some states, notably China and the US, have resources to train custom, high performance AI models with the capacity to perform influence campaigns on a global scale, autonomously exploit IT systems, or assist with development of new weapon systems. It would be naive to dismiss this as a motivating factor in the global AI race.

From Fear to Foresight: A Path Forward

The global AI race is in full swing, with scenes dominated by two countries that seek superiority, believing that advantage in AI will lead to competitive edge in areas like the economy, education, or the military. Even official documents (e.g., OpenAI's letter from March

2025) bring a well-known narrative of "Democratic AI" versus "Communist AI" and Donald Trump's executive order from January 2025 points to "America's global dominance."

In my opinion, in the long term, it may pay off to be smarter, not necessarily the biggest. Countries like India or Europe find their own strategy and niches. Europe, apart from building AI independence, focuses on de-risking, safety, and fundamental rights. While the regulation-based approach is well balanced and effective in restricting abuse of data of EU citizens, as well as safeguarding the riskier applications of AI, it will not stop malicious actors who operate lawlessly anyway. That's why the EU states must cooperate to detect, block, and respond to AI-powered influence campaigns or IT system exploitation to conduct espionage or sabotage operations.

What should you do? On a personal level, take advantage of opt-outs to protect privacy and engage in critical thinking as disinformation distorts our digital reality. Businesses have to be responsible when integrating AI and consider the risk of the full digital supply chain. On a government level, what matters is smart regulation that effectively de-risks the legally operating AI applications. Digital sovereignty is crucial on business and state levels. If that is not feasible, at least pursue diversification.

With the crime and nation state hacking risk, AI versus AI digital battles are coming. The only way to effectively stop AI threat actors is to build AI-based defense systems and cross-country cooperation in threat intelligence sharing to combat intrusions and disinformation campaigns and to track emerging threats. With the proper level of international cooperation, information sharing, and the participation of private sector companies, a 360-degree approach to AI safety is possible and viable.

About the Author

Leszek Tasiemski is a cybersecurity and AI risk expert with over 20 years of experience in offensive and defensive security. Leszek is employed at WithSecure, a European cybersecurity provider, as VP

R&D of Elements Cloud, advancing AI in cyber defense. Leszek promotes responsible AI and cybersecurity practices across the EU through GRAI at the Polish Ministry of Digital Affairs and by contributing to the European Commission's Plenary on the Code of Practice for General-Purpose AI. He is also dedicated to combating cyber violence, volunteering at NGOs to support victims and raise awareness. Leszek holds degrees and an MBA in computer science and economics and shares his expertise as a lecturer and university council member.

LinkedIn: https://www.linkedin.com/in/tasiemski/

CHAPTER 28

FROM THE HOLOCENE TO THE AI HORIZON

By James Taylor
AI & Data Strategy Leader
Royal Leamington Spa, United Kingdom

> *Study the past if you define the future.*
> —Confucius

The First Fire

The ice pulled back. The world warmed. People stepped out of caves and shelters of bone and hid, feeling a gentler sun on their faces. They did not know the word "holocene." They only knew that winter had loosened its grip and the air no longer bit so hard.

Fires burnt in many places: on the Siberian steppe, in the forests of the East, and on the grasslands of Africa. Men and women bent over the same work everywhere. Keep the fire alive. Keep the children warm. Keep hunger and fear at a distance. They fed the flames with brush and bone and learnt, by trial and pain, how close they could press their hands toward the flame before their skin burnt.

They watched the seasons turn and the stars wheel across the sky. They counted the days between migrations, learnt about river moods, and listened to storms. At night, they told stories, which contained what they had learnt. Which roots healed? Which clouds meant danger? Which winds would bring the herds back? They shared small tricks for staying alive in a complex world: how to read tracks, how to follow birds, and how to listen for ice breaking on distant water. The tales mixed truth, fear, and hope, but they kept people alive.

They lived close to the edge of life, close enough to feel its breath. Every choice mattered. A missed track meant hunger. A wrong story meant death.

But even then, they looked beyond survival. They drew lines in the dirt, shaped figures from clay, and traced the arc of stars with their fingers. They wanted more than safety.

They wanted understanding. A mind stretched by curiosity never returns to its old shape. That was true then, and it is true now.

Every age carries a question at its centre, and ours is no different. We try to understand what we are building and what it will make of us.

This event was the beginning of reason. They did not call it that. They only watched and remembered and tried not to die.

Knowledge grew slowly. A sharper stone appeared. The pot became steadier and more stable. The wheel began to roll farther. For years, the gains were small. Then the pace began to change. Villages grew where the soil was kind. Crops grew in rows instead of by chance. People had time to look up and wonder why the world worked the way it did.

Writing appeared. Marks on clay meant grain, water, and debt. Clay tablets carried memory beyond any single skull. A contract could outlive the hands that formed it. The wheel spread along the roads of early kingdoms. Carts and chariots followed lines worn into the earth.

Yet fire was older than all of it by hundreds of thousands of years. Fire made complex tools possible. Fire-hardened bricks. Fire-bent metal. Fire drove back in the dark. Fire shaped the minds that would one day shape machines.

Now, in our time, we sit before another kind of flame. It has no smoke or smell. It hums in servers and shines in screens. It answers questions in the dark and sweeps through oceans of data without sleep. This new fire is artificial intelligence. It is young, but the power in it feels old, like that first blaze on a cold plain when the night finally gave way. We are once again gathered around a light we do not yet fully understand.

Signs, Stories, and the First Machines

Long before anyone wrote poetry, people scratched lines on bone to count days. A hunter notched each kill. A trader pressed small clay tokens—cones, spheres, and little shapes—into jars to track his goods. Over time, those marks became cuneiform. Numbers turned into words. Accounting turned into law, prayer, and myth.

Farmers along the Nile carved gauges into stone and read the height of the river like a sentence. A few extra marks meant feast; too few meant hunger. In the Andes, runners carried quipus—ropes with knots tied in careful places—that held tallies and messages. A set of cords might remember a harvest, a tax, or a census. People in China used fire to break open tortoise shells. The cracks were read as signs. The signs became characters. Language stepped out of the flame.

The habit spread in many forms but with the same aim: watch, mark, predict. Tools became partners in thought. The abacus let fingers race ahead of speech. Merchants trusted the feel of beads sliding on rods more than their memory. A bronze box from a sunken ship— the Antikythera device—turned into a forest of gears—modelled the sky. Someone had built a mechanical heaven inside a metal case and trusted it more than the eye.

None of these things was alive. None could think. But each held memory beyond any single life. They were the first bridges from firelight to logic, from fleeting observation to structured record.

A machine does not dream, yet it can hold the dreamer's marks. It does not feel cold, yet it remembers winters.

Even then, people sensed a boundary forming, a line between thought and the tools that sustained it. They did not know the word "memory" as we use it now, but they felt the strangeness of seeing their past held outside themselves. It was the first hint that the mind might one day stretch beyond the skull.

People also dreamed of machines that moved on their own. The Greeks told of Hephaestus, the lame god at the forge, who hammered out golden servants and bronze giants. Talos circled Crete with molten metal in his veins. In Indian and Chinese tales, statues walked and chariots flew. Kings listened to mechanical men that bowed and spoke. In Baghdad, the Banu Musa brothers drew plans for fountains and instruments that worked with hidden valves. Al-Jazari built clocks and musical automatons that moved to the pull of water and weight and seemed almost to possess a will.

Across the world, practical genius matched the myths. Maya astronomers carved calendars into stone to track the sun and moon with startling precision. In Africa, the city of Meroë glowed with iron smelting. Its tools crossed deserts on the backs of caravans. In India, Pāṇini wrote a grammar of Sanskrit, so exact that it reads like code. At the House of Wisdom in Baghdad, scholars translated, debated, and refined. From the name al-Khwarizmi came the word algorithm.

Greek geometers drew proofs from lines and shadows. Surveyors along the Nile and in Mesopotamia measured fields erased by floods and marked new borders with rope and stake. In every land, people tried to make sense of the world with whatever lay at hand: bone, clay, metal, sand, water, light.

Wonder bows to the method. Myths bent toward models. The world became something that could be measured and reasoned. This marked the beginning of an era characterised by the emergence of silicon minds.

Civilizations rose and fell, but the hunger for order endured. People carved laws in stone, painted warnings on walls, and carried stories across mountains and seas. They built libraries that burnt down and then rebuilt them.

Each era added a layer to the same old impulse: to make knowledge last longer than the body. For thousands of years, this work moved slowly, tied to hand and breath.

Then came a force that would quicken the world beyond anything the old scribes imagined.

Ink, Iron, and the Birth of Code

Empires rose and fell. The hunger to know did not.

The printing press changed everything. Metal letters locked in frames pressed words onto paper. Pages travelled by river, road, and ship. Knowledge slipped from monasteries to markets and homes. One press could correct a mistake; the next could fix it everywhere. Truth, or something near it, could be copied. Ink carried diagrams of the body, sketches of machines, and charts of stars. Thought gained structure: tables, columns, numbers. Memory moved from the tongue to the page. A book could cross borders and centuries without losing its voice. That had never happened on such a scale.

Ideas marched across borders like armies. A page could travel farther than its author ever would. Knowledge grew restless. It no longer belonged to the gatekeepers but to anyone willing to read. It changed how people argued, prayed, and dreamed.

New tools followed. Glass lenses brought the moon close enough for Galileo to see its scars. Pendulums swung honest seconds into clocks. Napier's bones let merchants multiply. Pascal constructed a gear-driven device to assist his father in tax calculations, demonstrating that metal could grasp arithmetic.

Then the low thunder of Jacquard's looms rolled across Europe. Punched cards told threads when to rise or fall. Patterns once held by memory lived in holes in pasteboard. The cloth remembered. The loom became a script that could be changed without rebuilding the machine.

They learnt that any pattern could be unravelled and rewoven: text, melody, or code. Charles Babbage saw those cards and imagined something beyond cloth. He sketched a machine of cogs and levers that could compute tables. Ada Lovelace read his plans and envisioned even greater possibilities. The engine, she wrote, might weave algebra the way the loom wove silk. One day, she believed that such a machine might handle not only numbers but any symbol that encompasses music, images, and the fundamental elements of reason itself.

Their engine was never built. But the idea escaped the workshop. A machine could store instructions. A machine could follow them. A machine could, in its own way, think.

Descartes wrote that the human body was governed by gears and nerves. The mind, he said, was another flame. Leibniz imagined a language of pure logic. Boole turned thought into algebra. None of them saw a circuit board. But their ideas became its grammar.

Step by step, magic seeped from the gaps in knowledge. What once belonged to the gods moved under the lamp of experimentation. The world became a rule-bound machine. The rules can be learnt. And what could be learnt could be built.

War, Wires, and a World of Patterns

War has propelled the world forward repeatedly.

During the Second World War, Alan Turing sat in cold rooms under dim lights. He worked with relays, wires, and punched tape, machines built to search through cyphers faster than any human. His work helped break enemy codes. The devices he and his team built proved that a machine could solve problems too vast for the human hand.

After the war, he asked a tricky question: Could a machine ever think like a person? Could it speak convincingly enough to fool us?

He imagined a test, a simple exchange of typed words between a human and an unknown partner. If you could not tell which it was, had the machine begun to think?

At the same time, Norbert Wiener wrote of cybernetics, the study of feedback and control in animals and devices. Claude Shannon reduced every message to bits—1s and 0s—and showed how to send them through noise without losing meaning. John von Neumann designed a computer whose memory held both data and instructions. Almost every machine since has followed his pattern.

Vacuum tubes gave way to transistors, then to chips. Machines that once filled rooms collapsed into pockets and wrists. The speed at which thoughts once travelled through parchment has now accelerated to that of light.

In 1989, Tim Berners-Lee made the first web page. He linked documents, so scientists could share notes. Although creating the first web page was small in itself, its impact was significant. The first page was plain. There were no crowds, no cheers. But something immense shifted.

A map of the world's thought had begun, a place where memory had no weight, where distance could no longer divide ideas. For the first time, a person could speak across oceans without leaving a room. The fabric of humanity tightened, thread by thread, connection by connection. We had woven a web, and, slowly, the web began to weave us. The world's knowledge seeped into glass screens. The planet grew a nervous system of cables, routers, and satellites. The old spark glowed in server farms, street cabinets, pockets, and palms. We were ready, though we did not know it, for a new kind of mind.

A World Run on Patterns

After millennia of enhancing our physical abilities, such as our hands and sight, as well as our memory, we have created something that enhances the very process of thinking.

Artificial intelligence did not arrive with one big moment. It slipped into place. Initially, it was developed in laboratories. Next, it found its way into factories. Eventually, it became a part of our daily lives.

In China, a man pays with nothing but his face. Robots in Japan pick up the assembly line rhythm. In Europe, travellers hear their own language whispered while someone else's mouth speaks. In Africa, farmers ask questions in their tongue and receive advice on soil and seed. In Latin America, ride-sharing apps identify paths through chaos no map could show. In hospitals across the world, surgeons guide tools steadied by code.

Your phone suggests your next word, your next song. Your screens attempt to anticipate you. Your home learns your habits and shifts the heat. Each day, your home makes tens of billions of tiny decisions that are quiet and invisible to you. Often, we hardly notice. But in the still of night or a pause on a journey, we feel it: a silent helper sits in our hands.

Behind this convenience lies competition. Companies pour money into AI. Governments speak of advantage. Markets swing on the promise of more intelligent machines. Some refer to it as a new arms race. Some call it the next industrial revolution. It may be both.

Yet for all its power, AI has limits. It sees patterns in data. It predicts what might come next. But it does not understand as a human does. It has no morality, only the goals we assign it. It is a tool, not a soul.

Still, we lean toward our creations, listening for intention in their answers and meaning in their silences. We search their voices for intent, their silence for meaning. We have always filled the unknown with imagined life, faces in the fire and warnings in the wind. So when a machine answers in calm, perfect sentences, something in us responds. This response is not a result of the machine's feelings. But because we do.

A tool shaped without care can cut the hand that holds it, a truth as old as the first blade. Still, AI reaches far.

In India, a farmer snaps a picture of a sick leaf. In Australia, satellites sense the heat of a fire before the forest knows it is burning. In Brazil, people without credit histories secure loans because a model shows promise. In classrooms, lessons shift to meet each child's needs. In observatories, AI searches starlight for patterns human eyes miss. The machine has taken root in many lands. It carries local accents and needs. It belongs to no one and touches everyone.

Guardrails and the Next Dawn

But a tool powerful enough to see what we can't also carries the weight of what we choose to ignore. As this power spreads, the world looks for a line it will not cross. Nations and councils draft principles. They speak of dignity, rights, fairness, and transparency. They argue that AI must remain under meaningful human control. Laws emerge to govern high-risk systems. Some regions insist on informing people when a machine makes a crucial decision regarding their lives. Others push for an AI Bill of Rights, not for the code, but for the citizens who must live beside it.

The same questions keep arising:

- How do we keep data private in a world built on sharing?
- How do we stop algorithms from repeating old prejudices at scale?
- How do we help workers displaced by automation?
- How do we teach children to use these tools without dulling their curiosity?

A simple truth sits beneath all of this: AI amplifies what we already are. If our society is unfair, the machine will learn that injustice and repeat it. If we strive for wisdom and mercy, it can help carry those values farther.

The tools of tomorrow will be even more advanced. How do we teach children…? We are developing quantum machines that surpass the boundaries of today's technology. Hybrid computers grow from neurones. Neural interfaces facilitate the transmission of thoughts through circuits. Autonomous probes traverse far-flung worlds. Models are actively exploring new avenues in climate, medicine, and energy.

Every generation encounters a pivotal moment. Some rise to meet it. Some turn away. But no age escapes the wager of progress, the bet placed on our wisdom. If we build without conscience, we inherit chaos. If we build with care, we inherit possibility.

The future is not a straight road but a forked one, lit only by the hope we choose to carry. Each promise carries risk. Power without principle leads to ruin. Intelligence without conscience becomes cunning. Technology without humanity is a cage.

But tools shaped with care become extensions of the hand and mind. Machines guided by ethics can widen our reach. Knowledge tied to wisdom can light the way rather than burn it down.

We stand between two fires: The first spark our ancestors coaxed from cold tinder. The second is the glowing servers and screens that surround us now.

The story of AI is, in the end, our story, the story of a species that learns to watch, remember, imagine, and finally build something that reflects its own thinking. The question is not only what these systems will become, but what we will become beside them. If we remember that every line of code comes from human hands, then we remember who holds the flame.

In the end, the fire will remember only the hands that held it, and the future will rise or fall by the strength of that grip. And so we return to the truth that has followed us since the first fire. And like the first fire, this force will shape us long before we learn to control it.

Intelligence is the ability to adapt to change.

—Stephen Hawking

References

1. Stephen Hawking, *Brief Answers to the Big Questions* (London: Bantam Press, 2018) —quotation "Intelligence is the ability to adapt to change."

2. Confucius, *Analects*, bk. 2, sec. 11—"Study the past if you would define the future."

3. World Intellectual Property Organization (WIPO), *Generative AI: 2025 Patent Landscape Report* (Geneva: WIPO, October 2025)—global AI patent and investment data.

4. United Nations Educational, Scientific and Cultural Organization (UNESCO), *Recommendation on the Ethics of Artificial Intelligence* (Paris: UNESCO, 2021)—adopted by 193 member states; core principles of fairness, transparency, and human oversight.

5. Organisation for Economic Co-operation and Development (OECD), *OECD Principles on Artificial Intelligence* (Paris: OECD, 2019; reaffirmed 2024)—human-centred AI policy framework.

6. African Union Commission, *Continental Strategy for Artificial Intelligence* (Addis Ababa: African Union, 2024)—includes an "Ubuntu" framework for ethical AI.

7. Reuters Business Desk, "Nvidia Becomes First $5 Trillion Company as AI Boom Redefines Tech Sector," October 29, 2025.

8. Goldman Sachs Research, *The AI Investment Wave: Infrastructure and Productivity Outlook* (New York: Goldman Sachs, 2025).

9. Lawrence Wong, address to the Asia-Pacific Economic Cooperation (APEC) Summit, Honolulu, November 3, 2025—remarks on AI training and inclusive growth.

10. Interpol Digital-Forensics Symposium, Tokyo, October 28–30, 2025, presentation of the AlchemiX Deepfake Detection System by Singapore GovTech and the Home Team Science & Technology Agency.

11. Michio Kaku, interview by Piers Morgan, *Uncensored*, TalkTV, November 2025; and Michio Kaku, *The Future of the Mind: The Scientific Quest to Understand, Enhance and Empower the Mind* (London: Penguin, 2014).

12. Norbert Wiener, *Cybernetics: Or Control and Communication in the Animal and the Machine* (Cambridge, MA: MIT Press, 1948).

13. Claude E. Shannon, "A Mathematical Theory of Communication," *Bell System Technical Journal* 27 (1948): 379–423, 623–56.

14. Ada Lovelace, "Notes on the Analytical Engine," in *Scientific Memoirs*, no. 3 (London, 1843) —prediction of machines creating music and art.

15. Gottfried Wilhelm Leibniz, *De Arte Combinatoria* (Leipzig, 1666)—binary mathematics and universal language concept.

16. Alan M. Turing, "Computing Machinery and Intelligence," *Mind* 59 (1950): 433–60—the Imitation Game (Turing Test).

17. Banū Mūsā brothers, *Book of Ingenious Devices* (Baghdad, ca. 850 CE)—earliest record of programmable automata.

18. Ismail ibn al-Razzaz al-Jazari, *The Book of Knowledge of Ingenious Mechanical Devices* (1206 CE)—detailed mechanical designs for robotic musicians and clocks.

19. Filippo Brunelleschi, architectural drawings and experiments on linear perspective (ca. 1415), as recorded by Antonio Manetti in *Vita di Filippo di Ser Brunellesco*.

20. Pāṇini, *Aṣṭādhyāyī* (ca. 5th century BCE).

21. Vaclav Smil, *Energy and Civilization: A History* (Cambridge, MA: MIT Press, 2017).

About the Author

James Taylor is a recognized authority on artificial intelligence, data strategies, and ethical innovation. He leads complex data transformations across financial services, global retail, and government. His focus is simple: turning AI from a buzzword into real-world impact—with products that are trusted, adopted, and built to last.

James blends deep technical fluency with sharp strategic vision. He has delivered AI-enabled platforms across both public and private sectors, from financial services and agri-food to Fortune 500 giants. Whether shaping enterprise data lakehouses or co-designing predictive models for ESG performance, his work bridges engineering, ethics, and practical adoption.

A published author and international speaker, James is co-authoring a landmark book on AI and society. He is also the founder of www.taylortailored.co.uk, where he shares tools, insights, and experiments at the intersection of intelligence and innovation.

CHAPTER 29

THE HUMAN OPERATING SYSTEM: HOW ORGANIZATIONS ARE SABOTAGING THEIR AI STRATEGY

By Megan Thomas
Change & Communication Leader, AI Educator
Sydney, Australia

If you're one of the many business leaders investing millions in AI and wondering why results aren't showing up, you're not alone. Only 5% are seeing returns.[1]

Your tech is in place. Policies are written. Most employees are already using AI. We know the technology *can* deliver value. So why, instead of innovation and productivity nirvana, are organizations hitting a wall? Because it doesn't matter how good your system engineers, data, and platforms are, they have little to do with realizing your bottom-line gains.

AI Technology Is Not Failing. The System Around It Is

AI fundamentally changes how decisions are made, how organizations operate, the risks they're exposed to, and how work gets done every day.

Most employees are already using AI at work, whether they're allowed to or not, and they're doing it with no role-specific direction, unclear policy, and weak data governance. At the same time, managers are expected to lead AI adoption, organizational structures need to be redesigned, and no one can clearly explain how roles will change. Leaders are under pressure to deliver, and everyone is filling the gaps because the organization simply can't keep up. In summary: Organizations are preparing for AI in the most superficial way, if at all. It's like revving a car with no gas, no roads, no road rules or map, then wondering why it's not going anywhere.

We don't have to look far to see what happens next. A Deloitte consultant left AI-generated errors in a client report.[2] A finance employee at Arup was tricked by a deepfake into handing over millions.[3] Companies like Amazon and Safetrac face legal action over intrusive worker surveillance.[4,5] These are human system failures.

AI is not a technology upgrade. It is a strategic operational and cultural change that's vastly underestimated. However, change management is one of the most controllable drivers of ROI.

This isn't consultant theory. It's based on more than 20 years of practical experience working inside organizations, alongside executives and their teams. What follows is the reality of what's happening and what leaders must do differently to build profitable, sustainable businesses in an AI world.

The 10 Ways Organizations Sabotage Their AI Strategy

Here are ten ways organizations are ignoring the human operating system and sabotaging their own AI strategies.

1. Leadership and strategy are missing.

In the rush to get AI "out there", strategic direction has been sidelined and replaced with vague expectations of "efficiency gains." Boards are concerned about AI risk, but most board members have "limited to no knowledge or experience" with AI, and for a third of boards it's not even on the agenda.[6]

Technology initiatives are still treated as IT projects, with budget, resources and decision rights following. The system gets "delivered" but isn't fit for purpose. The "fail fast" methodology borrowed from tech startups was supposed to make organizations more adaptable but is reckless when applied to systems that affect people's livelihoods, safety, and trust.

Earlier in my career, I was working on crisis management for a payroll system fiasco at Queensland Health.[7] Rolled out too early, 80,000 doctors and nurses were paid incorrectly and taxpayers also paid the price: around US$1.4 billion in today's dollars. Headline news for years and subject of many case studies, the health commissioner's postmortem described it as one of the worst failures in Australian public administration.

A decade later, the disastrous Robodebt scheme even eclipsed this.[8] Automated systems wrongly issued debts to welfare recipients, causing immense harm, legal action, and compensation payouts. Different technology, same underlying failure.

What happened in both cases is repeated every day on a smaller scale across industries and countries. On-time, on-budget delivery is prioritized over everything else. Requirements are poorly understood. People impacts aren't assessed. IT is rewarded for delivery, while the business absorbs the fallout in workaround costs, lost productivity, and unnecessary stress.

This approach is no longer survivable with AI. The stakes are higher, the impact broader, and the consequences faster.

2. Stakeholders are brought in too late.

If you are designing, building, or deploying AI systems without considering your stakeholders (employees, customers, and others), you are putting the organization at risk. Too often, project leaders dismiss engagement altogether: "We'll never get agreement," "It will slow us down," or "We sent an email." Communication, if it happens at all, is a list of features rather than clarity on what is changing, why it matters, and who it affects.

Of course, it's not possible to meet everyone's needs, and decisions must be made. Consulting early, managing expectations, and making course corrections upfront is far less costly than dealing with resistance, rework and reputational damage later.

3. Confidential data is walking out the door.

More than half of workers hide their AI use from their employer, and nearly half admit they have uploaded company data into public AI tools. [9] At global scale, that means millions of employees sharing confidential information with the public and competitors.

Some organizations respond by blocking access to AI tools altogether. With "shadow AI" already widespread, this approach doesn't stop it being used. It just removes visibility, control, and skill-building opportunities.

We've seen it play out before. Organizations blocked access when Facebook first arrived, preventing employees from seeing customer feedback and creating issues leaders didn't know existed. AI is no different. If leaders don't provide safe, clear, and practical ways for people to use it, employees will make their own decisions and risk multiplies.

4. People skills are undervalued.

According to the World Economic Forum (WEF), by 2030, around 40% of today's skills will be outdated.[10] The most in-demand skill globally will be analytical thinking. In the top five fastest growing skills are creative thinking, leadership and influence, resilience, and flexibility.

These "soft skills" are paid lip service but still dismissed as secondary to technical expertise. AI has turned that thinking into a commercial liability. Even cybercriminals design their systems around psychology and social engineering, because they understand human behavior is often the weakest link.

Today, technologists must be really good business communicators, business leaders must be data literate, and HR must be financially literate. Everyone in every role must broaden. Those who can bridge technical, business, and human skills will be the most effective.

5. Training isn't building real capability.

Organizations say training is a top priority. In reality, 70% of employees received no AI training in the past year.[11] Where training does exist, it's often a handful of webinars and instruction to "experiment," followed by an assumption AI has been "adopted." It hasn't.

Instead of providing role-specific AI skills for new ways of working, people are getting generic AI training or a one-off workshop, leaving employees with little confidence they're building the right skills.

For organizations unwilling to invest in proper training, there is a brutal skills shortage coming. If people feel like they're falling behind, they won't be waiting around, they will leave. AI experts are predicting the first billion-dollar companies built by a single worker are close, [12] so your talent won't just move to competitors; it will be able to do something completely new.

6. Work is not being redesigned for AI.

Say hello to your new digital coworkers, agentic AI. They can plan, act, and learn autonomously. As these agents get better, human work will move further away from task execution and toward AI orchestration and oversight.

You can't just sprinkle AI onto existing tasks and expect better outcomes — AI can make poor processes worse. Jobs should be redesigned end-to-end, clearly separating tasks suited for automation from those requiring human judgment. There will also be new responsibilities for managing both people and AI agents. Performance measures will need to change too, reflecting not just individual output, but how effectively someone works with AI.

7. Leaders overestimate how well they communicate.

Politics and power dynamics make it hard for executives to see their own blind spots. Research [13] tells us 99% of business leaders believe they communicate change well, but only 25% of employees agree. Just 12% to 16% rate the critical updates they receive as "very effective".

This means goals are misunderstood, time is wasted clarifying and correcting work, projects fail, morale drops, and customers are left confused or dissatisfied.

AI can help draft messages faster, but it will not fix this problem. Leadership communication must be active, visible, and credible.

8. Employee trust is breaking down.

Trust in institutions has been eroding for years. AI can make it worse. Alongside cost-cutting, poorly handled restructures and return-to-office mandates, mistrust is rampant. The term "AI anxiety" is now appearing too.

Ask employees what worries them most about AI and you'll hear:

- *"Is my job safe?"*
- *"How will my role change?"*
- *"Are they monitoring what I'm doing?"*
- *"Who's responsible if AI gets it wrong?"*

This isn't resistance to change. It's a normal response to uncertainty. Employees expect leaders to be honest, to listen, and to see their actions match their words.

Executives often don't realize how much power employees hold. The Hollywood writers' strikes showed how quickly AI can become a bargaining weapon.[14] Even OpenAI's board discovered they couldn't remove their CEO without 90% of their staff threatening to quit. [15] When trust breaks down, employees can undermine strategies, sabotage systems, leak information, and damage reputations.

If people trust how AI is being used, adoption is faster, and there are greater returns. One interesting example is China. Trust in AI is high, and they are seeing more benefits than many other countries.[16]

You can't force AI adoption, you have to build trust first. Your people will decide whether your AI strategy succeeds or fails, so be sure you're not creating enemies on the inside.

9. Output increases while quality declines.

AI is making some tasks faster, but quality is taking a nosedive. Alarmingly, 66% of employees say they do not verify the accuracy of AI outputs.[17]

AI "work-slop" is draining productivity and damages credibility. The cost is estimated at around US$186 per employee per month, (I'd argue at least), and recipients rate the sender as less competent.[18]

I remember gasping when 2023 reports said up to 90% of online content could be synthetically generated by 2025.[19] Looking back, the headline was hyped, but current estimates put it well over 50% and growing fast. It's not just the overwhelming volume of poor-quality content we should be worried about. Cognitive load and attention spans stretched to the max mean we are even more exposed to misinformation, scams and manipulation.

These days it only takes a handful of images and seconds of audio to create a convincing deepfake. Imagine a damaging video of your CEO circulating online, amplified by bots, triggering panic among employees, customers, and investors.

If this happens, your employees are your first line of defense. Most are not equipped to tell fact from fiction before reputational damage and a plummeting share price take hold.

10. Regulation is not the answer.

Regulation is uneven and, in most regions, inadequate. Planning to remediate only when something goes wrong is a dangerous strategy. With AI, something will go wrong.

Transparency about how AI is used is a minimum standard leaders can set now. It shouldn't require regulation to explain when AI is involved in decisions that affect employees, customers, or the public.

Reports vary, but less than half of organizations have a formal policy or guidelines in place.[20] Where they do exist, they are written in legalese with little practical guidance. Communication typically amounts to an email with a policy link.

The Illusion of AI Adoption

Now we've looked at the ten ways organisations are sabotaging their AI strategies by ignoring the human operating system. The final nail in the coffin is a lack of awareness of these issues and of what true

AI adoption — the kind that delivers a return on investment — really means.

As the saying goes, "What gets measured gets done", but in this case we're measuring the wrong thing. Sure, there are several well regarded "AI maturity indexes", but they don't measure how people are actually using AI, what's changed in roles and workflows, what capability has been built, or whether any real value is being created. Having strong technical or legal foundations doesn't mean you have adopted AI.

If "AI adoption" is measured, it usually means technical deployments or use cases. In other words, "we launched an IT project." The number of people trained is sometimes used as a proxy, but volume tells you nothing about relevance or outcomes.

The same illusion exists when the "AI talent" metric (number of AI resources hired), is used to show strong investment in people related AI capability. Armies are hired to build AI systems, but often you won't find a single resource dedicated to AI change management, communication, or future workforce planning. These necessities are usually pushed onto existing teams without the capacity, influence, or budget.

As a result, AI adoption is a matter of perception, polished in annual reports and PR. The one thing that actually proves success—people working differently to deliver business outcomes, remains largely invisible.

Don't be fooled.

Deploying models ≠ adoption.

Using AI ≠ adoption.

Number of AI Engineers ≠ adoption.

Stop sabotaging your AI strategy and start turning your investment into impact.

AI is already part of how your organization works. The question is whether it's creating value or risk. Rolling out technology, running training sessions, and publishing policies will keep you busy, but it won't deliver returns.

This isn't a problem to be solved by better tools or tighter regulation. It's a leadership issue that can be addressed now. Organizations that put people first in their AI decisions move faster, avoid costly mistakes, and get better results. Those that don't will keep spending and wondering why nothing changes.

A note on my AI use: I use AI to help with research and editing content for clarity. The thinking, experiences, structure, and phrasing are all mine.

Stop sabotaging your AI strategy by investing in the human operating system. Get *The Human Operating System: AI Checklist for Leaders* © along with more practical tips. Visit www.buzzcomms. com.au.

THE HUMAN OPERATING SYSTEM: AI CHECKLIST FOR LEADERS©

If you can't tick most of these, ROI risk is high:

Leadership and Strategy

☐ CEO leads the AI strategy as a business priority, not a technology initiative.

☐ Success is measured beyond cost savings, including quality, risk, and reputation.

☐ Executives are accountable for outcomes, not just delivery milestones.

Governance and Risk Management

☐ AI governance covers strategy, ethics, legal, risk, and stakeholder engagement.

☐ People impacts are fully assessed; skills, workload, roles, wellbeing.

☐ Strong guardrails are in place to prevent unethical use and harm.

☐ AI policies are clear and understood by employees.

☐ Escalation paths exist for errors, bias, safety, or ethical concerns.

Communication and Culture

☐ Change Management has active, executive sponsorship and budget.

☐ Leaders are confident and clear on the AI strategy and objectives.

☐ Day-to-day work impacts are understood.

☐ All surveillance is clearly explained.

☐ AI reduces barriers for language, disability, and cognitive load.

☐ Employees are involved early, not just at rollout.

Work Design and Skills

☐ Roles are redesigned end-to-end, rather than adding AI on top.

☐ Managers are equipped to coach and lead AI use.

☐ Training is role-specific and ongoing.

☐ Employees know when to use AI, when not to, and how to challenge it.

Most importantly, don't forget to unplug. Let's make AI work for people and purpose.

© 2025 Buzz Communications Australia

Endnotes

1. Massachusetts Institute of Technology (MIT), *State of AI in Business 2025* (Cambridge, MA: MIT Project NANDA, July 2025), https://nanda.media.mit.edu/ai_report_2025.pdf.

2. Nino Paoli, "Deloitte Was Caught Using AI in $290,000 Report to Help the Australian Government Crack Down on Welfare," *Fortune*, October, 7, 2025, https://fortune.com/2025/10/07/deloitte-ai-australia-government-report-hallucinations-technology-290000-refund/.

3. "Arup Lost $25 Million in Hong Kong Deepfake Video Conference Scam," *Financial Times*, May 17, 2024, https://www.ft.com/content/b977e8d4-664c-4ae4-8a8e-eb93bdf785ea.

4. Michael Sainato, "'You Feel Like You're in Prison': Workers Claim Amazon's Surveillance Violates Labor Law," *The Guardian*, May 21, 2024, https://www.theguardian.com/us-news/article/2024/may/21/amazon-surveillance-lawsuit-union.

5. "Company Turned Laptops into Covert Recording Devices to Monitor WFH," *Australian Financial Review*, August 24, 2025, https://www.afr.com/work-and-careers/workplace/company-turned-laptops-into-covert-recording-devices-to-monitor-wfh-20250822-p5mp0z.

6. Deloitte, *Governance of AI: A Critical Imperative for Today's Boards* (New, York: Deloitte Insights, 2025).

7. "2010 Queensland Health Payroll System Implementation," *Wikipedia*, last modified 2025, https://en.wikipedia.org/wiki/2010_Queensland_Health_payroll_system_implementation.

8. Australian Broadcasting Corporation, *"Robodebt Condemned as a 'Shameful Chapter' in Withering Assessment by Federal Court Judge"*, 2021, https://www.abc.net.au/news/2021-06-11/robodebt-condemned-by-federal-court-judge-as-shameful-chapter/100207674

9. "Employees Are Using AI More Than Their Bosses Realise, KPMG Study Finds," *Business Insider*, 2025, https://www.businessinsider.com/employees-using-ai-more-than-bosses-realise-kpmg-study-finds-2025-?.

10. World Economic Forum, *The Future of Jobs Report 2025* (Geneva: World Economic Forum, 2025), https://www.weforum.org/reports/the-future-of-jobs-report-2025.

11. Dayforce, *16th Annual Dayforce Pulse of Talent* (2025), https://www.dayforce.com/who-we-are/newsroom/16th-annual-dayforce-pulse-of-talent-71-of-workers-untrained-in-ai-as-adoption-and-expectations-su

12. "When Will AI Bring Us a Billion Dollar Start Up with a Single Worker?" *The Times*, 2025, https://www.thetimes.com/business/technology/article/when-will-ai-bring-us-a-billion-dollar-start-up-with-a-single-worker-qm90tf5m5.

13. Ragan Communications, "By the Numbers: Change Comms and Leadership," *Ragan Communications*, 2023, https://www.ragan.com/by-the-numbers-change-comms-and-leadership/.

14. Brookings Institution, "'Hollywood Writers Went on Strike to Protect Their Livelihoods from Generative AI,'" *Brookings*, 2023, https://www.brookings.edu/articles/hollywood-writers-strike-ai/. (Add access date.)

15. "OpenAI Staff Threaten to Quit Unless Board Resigns," *Wired*, 2023, https://www.wired.com/story/openai-staff-threaten-to-quit-unless-board-resigns/.

16. University of Melbourne and KPMG, *Trust, Attitudes and Use of Artificial Intelligence: A Global Study 2025* (Melbourne: University of Melbourne and KPMG, 2025), https://kpmg.com/xx/en/our-insights/ai-and-technology/trust-attitudes-and-use-of-ai.html. .

17. University of Melbourne and KPMG, *Trust, Attitudes and Use of Artificial Intelligence: A Global Study 2025*.

18. "AI Generated Workslop Is Destroying Productivity," *Harvard Business Review*, 2025, https://hbr.org/2025/?.

19. "Is AI Quietly Killing Itself—and the Internet?" *Forbes*, 2024, https://www.forbes.com.au/news/innovation/is-ai-quietly-killing-itself-and-the-internet/.

20. UST Survey Insights, *Navigating the Ethical Maze of AI Implementation* (2025), https://www.ust.com/en/insights/ust-survey-insights-navigating-the-ethical-maze-of-ai-implementation.

About the Author

Megan Thomas is a change management and communication leader specializing in AI. Globally recognized for her work, she has presented keynotes and published extensively on AI's impact on communication, leadership, and organizational change.

With more than 20 years' experience, Megan has helped senior executives lead complex people and technology transformations across media, professional services, and financial services, including within one of the world's largest AI operations. More recently she founded Buzz Communications Australia, where she consults, writes, and speaks on AI-driven change.

Email: megan@buzzcomms.com.au

Website: www.buzzcomms.com.au

LinkedIn: https://www.linkedin.com/in/meganthomas1/

PREDICTABLE VOLATILITY: WHY MEANING, NOT MODELS, DETERMINES WHETHER AI BREAKS YOU OR BUILDS YOU

By Andrew J. Turner
Advisor to Founders, Leadership Teams & Boards
London, England, United Kingdom

The single biggest problem in communication is the illusion that it has taken place.

—often attributed to George Bernard Shaw

On a cold October morning in 2010, Seema Misra opened her small post office branch in Surrey as she had done countless times before. She was eight weeks pregnant. Her young son stood beside her, tugging at her coat as she prepared for another ordinary day. Moments later, police officers walked in, placed her in handcuffs, and accused her of stealing £74,000.

Her alleged crime? A discrepancy generated by Horizon, the post office's centralized accounting system.

Inside the organization, Horizon had become something more than a tool. It had become a shield. If Horizon displayed a shortfall, a shortfall existed. And if a shortfall existed, someone must have taken the money. It was a perfectly circular logic, one that left no room for uncertainty, nuance, or human judgment.

Seema insisted she had taken nothing. Horizon insisted she had. Horizon won. She was sentenced to 15 months in prison for a crime that never happened.

Seema's case was not unique. More than 900 sub-postmasters were wrongfully prosecuted over two decades in the UK. Families were destroyed. Life savings were wiped out. Some did not survive the consequences. It now stands as one of the worst miscarriages of justice in British history.

Yet, to view this solely as a "glitch" is to miss the terrifying point. The software bugs in Horizon were fixable. The *meaning* failure in the post office executive suite was fatal. This tragedy reveals a lesson that applies to every organization adopting AI systems today: Catastrophic failures in the age of AI are not failures of technology. They are failures of meaning.

The Failure Hiding in Plain Sight

When people imagine AI failures, they picture faulty algorithms, rogue automation, or "hallucinating" chatbots. But the most dangerous and expensive failures, the ones that shatter trust and destabilize organizations, rarely come from technical glitches. They come from interpretive collapse.

Over 25+ years working with global enterprises and fast-growth startups, I've watched organizations break not because their models were wrong, but because their interpretation of those models was disconnected from reality.

- Meaning drift: When a definition slowly changes over time without anyone noticing
- Assumption creep: When a constraint in the model (e.g., "valid only for US markets") is forgotten as the tool scales
- Silent redefinitions: When "active user" means one thing to marketing and another to engineering
- Metrics used outside their intended purpose: When a rough estimate becomes a hard target

Humans can compensate for these gaps with intuition and experience. If a report looks wrong, a human manager asks, "Wait, does this include the holiday data?"

AI systems cannot do this. They possess neither intuition nor curiosity. They execute whatever meaning they have been fed, even if that meaning is flawed, incomplete, or inconsistent.

Meaning is not a philosophical abstraction. Meaning is operational infrastructure. When meaning collapses, trust collapses. When trust collapses, coherence disappears. And when coherence disappears, volatility becomes unmanageable.

The Inversion: The "Benevolent Human" Fallacy

To understand why meaning matters, we must apply inversion thinking. We must look at the counter argument. It is tempting to believe that humans are the "safety net" for AI, that if we just keep humans in the loop, they will provide the nuance and empathy the machine lacks.

The post office scandal proves this assumption false.

In that case, humans did not use their judgment to correct the machine; they used the machine to validate their own biases. The executives likely knew that "shortfall" was an ambiguous term. It could mean theft, it could mean error, or it could mean delay. But "theft" was a convenient meaning. It protected the brand. It shifted liability to the sub-postmasters.

The danger of AI is not just that it lacks meaning. The danger is that it provides a veneer of mathematical objectivity that allows humans to hide their own corrupt or lazy meanings. When a leader says, "The algorithm decided," what they are often saying is, "I have chosen a definition of success that I don't want to defend personally." Catastrophic failure happens when AI scales a corrupt human interpretation beyond the point of reversibility.

The Cycle of Resilience

Organizations under pressure often respond to volatility by demanding more precision, more dashboards, more KPIs, more automation. Ironically, this usually increases fragility rather than reducing it.

In the post office, leadership trusted Horizon's digital numbers more than the physical cash counts in their employees' hands. They traded reality for the illusion of accuracy. After watching this pattern repeat across dozens of companies, I developed the Predictable Volatility loop, a model for understanding how organizations remain stable even when volatility rises.

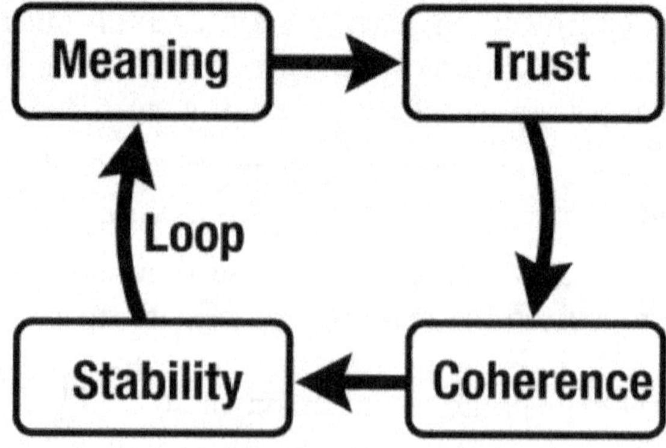

Figure 1: The Predictable Volatility Loop: a four-stage resilience cycle showing how meaning, trust, coherence, and stability reinforce one another. Source: Author.

Let's break down the mechanics of this loop, because if one link breaks, the system fails.

1. Meaning (The Anchor): Meaning is the agreement on what reality is. It requires clear ontologies. What is a "customer"? What is "churn"? What is "risk"? If these aren't defined explicitly, the model is hallucinating from day one.

2. Trust (The Connector): Trust is not an emotion; it is a calculation. It arises when humans believe the system reflects the reality defined in stage one. If a dashboard says sales are up, but the bank account is empty, trust evaporates. Without trust, users will either ignore the AI (shadow IT) or blindly follow it into a cliff (automation bias).

3. Coherence (The Action): Coherence happens when humans and machines interpret signals the same way. When the AI flags a "high-risk transaction," does the compliance officer treat it as a lead to investigate or a verdict to prosecute? Coherence is shared intent.

4. Stability (The Result): Stability is not the absence of change; it is the ability to withstand change. Stability becomes possible only after meaning, trust, and coherence are aligned.

The Feedback Loop: Crucially, stability allows you to circle back to meaning. When an organization is stable, it has the bandwidth to ask, "Are our definitions still correct? Has the market changed?"

Panic destroys meaning. Stability protects it.

When Misalignment Becomes "Truth"

Semantic drift is easy to dismiss because the language stays the same. "Risk" remains "risk." "Accuracy" remains "accuracy." But underneath familiar labels, meanings have changed.

Ask three executives to define "risk." Ask three analysts to define "accuracy." Ask three teams to define "customer value." You'll get incompatible definitions, all expressed confidently.

Humans can navigate this complexity through conversation. AI cannot. AI requires explicit instruction. Unless meaning is made explicit, the AI will lock onto a mathematical pattern that may have nothing to do with business reality.

That's why Horizon was so devastating. A shortfall did not mean: "There might be a system issue." It meant: "We have caught a thief." Once that meaning calcified, the system became impossible to question.

Another sub-postmaster, Parmod Kalia, borrowed £22,000 from his elderly mother to "repay" phantom shortfalls generated entirely by Horizon. When anomalies continued, he was prosecuted and sentenced to prison. The theft never happened. But the interpretation did.

Meaning Collapse in the Sky: The Boeing 737 MAX

If Horizon shows how meaning fails inside an accounting system, the Boeing 737 MAX shows how meaning failures can be physically fatal.

Between October 2018 and March 2019, multiple Boeing 737 MAX aircraft tragically crashed, killing 346 people. Investigations revealed that a software system called MCAS, the maneuvering characteristics augmentation system, played a central role. But MCAS did not malfunction in the traditional sense. It operated exactly as it was coded to operate. The catastrophe came from meaning divergence, a specific form of drift where different stakeholders hold contradictory definitions of the same system.

- To Boeing's engineers: MCAS meant a hidden, background stability-assist mechanism. It was a "helper" script designed to activate only in rare, specific aerodynamic conditions.

- To pilots: MCAS meant ... nothing. Because it was defined as a background helper, it was deemed so minor that it wasn't even highlighted in the flight manuals. They didn't know it existed.

- To regulators: MCAS meant an extension of an existing trim system, not a fundamentally new behavior that required recertification.

- In reality: MCAS meant a system that could repeatedly and aggressively override pilot input based on a single faulty sensor, forcing the nose of the plane down.

This gap between what MCAS *was* and what stakeholders *believed it was*, became catastrophic. Meaning drift had seeped into documentation, training, certification, and internal language. Everyone thought they were talking about the same thing. They were not.

The 737 MAX disasters reveal a truth about AI-era systems: When humans and machines assign different meanings to the same signal, control collapses.

The $569 Million Math Error? No, A Meaning Error.

In 2021, Zillow attempted to reinvent home-buying using its Zestimate model. The plan was "iBuying," using the algorithm to instantly buy homes, flip them, and sell them for a profit. When the initiative collapsed, the media blamed "the algorithm."

But let's look closer. The Zestimate algorithm had a published median error rate of roughly 6.9%. In the world of consumer curiosity, checking what your neighbor's house is worth, a 7% error margin is impressive. It's fun. It's useful. However, Zillow's business model for iBuying relied on razor-thin profit margins, often less than 5% per home after repairs and fees.

Do the math: You cannot build a business requiring 5% precision on top of a model with 7% variance.

The failure wasn't the model. The model was doing exactly what it was built to do: provide a rough estimate. The failure was the meaning. Internally, the meaning of the Zestimate shifted. It drifted from being a "probabilistic estimate" (a guess) to a "deterministic valuation" (a price tag).

- Probabilistic meaning: "There is a high chance this house sells for $400k."

- Deterministic meaning: "This house IS worth $400k."

When the housing market cooled, this semantic error became a financial disaster.

- $569 million write-down and approximately 2,000 layoffs.

- Complete shutdown of the division.

Meanwhile, competitors like Opendoor, using similar data but strictly controlling the meaning and application of their models (keeping humans in the loop for edge cases), survived. Zillow didn't suffer a model failure. Zillow suffered a meaning failure.

The Counter-Argument: The Cost of Rigor

I often hear pushback from startup leaders and agile product managers: "This sounds like bureaucracy. If we spend months debating the definition of 'churn,' we'll go out of business. We need to move fast."

This is a valid inversion. Rigidity is the enemy of velocity. If you apply heavy semantic governance to a music recommendation engine or a spell-checker, you are wasting money. The cost of a bad song recommendation is near zero. However, the cost of a bad hiring decision, a bad loan approval, or a bad flight control adjustment is non-linear. As Jeff Bezos has described, we must distinguish between type 1 decisions ((irreversible, high stakes) and type 2 decisions (reversible, low stakes).

- Type 1 (High Risk): Meaning must be rigid. Definitions must be locked.

- Type 2 (Low Risk): Let meaning be fluid. Let the AI experiment.

The Zillow failure happened because they treated a billion-dollar real estate portfolio like a type 2 decision. The Boeing failure happened because they treated a critical safety system like a minor software patch.

You don't need semantic rigor for everything. But you need it for the things that can kill you.

The Meaning Mesh: A Semantic Trust Layer

So, how do we operationalize this? We need a framework that sits between the raw data and the human decision-maker. I call this the Meaning Mesh.

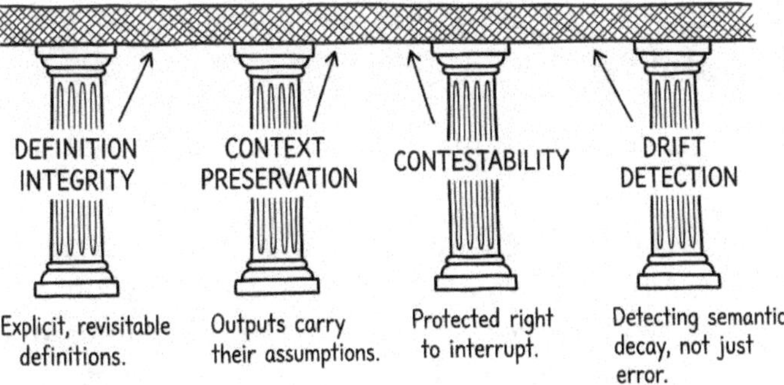

Figure 2. The Meaning Mesh - a four-pillar semantic governance framework for preventing meaning collapse in AI-driven systems.
Source: Author

The Meaning Mesh is not software; it is a governance protocol. It consists of four non-negotiable design obligations:

1. Definition Integrity

The Rule: One term. One meaning. No exceptions.

The Practice: You cannot have Sales defining "Revenue" as "booked contracts" while Finance defines it as "money in the bank." The AI will not know the difference; it will optimize for the number it sees. You must enforce explicit, revisitable definitions for every variable feeding the model.

2. Context Preservation

The Rule: Outputs without context are instructions without justification.

The Practice: Every model output must carry its "nutritional label." How sure is the model? Where did this data come from? Under what conditions is this output invalid? (e.g., "This Zestimate is valid for standard homes, not historic properties.")

3. Contestability

The Rule: If a decision cannot be challenged, it will eventually be wrong.

The Practice: There must be a protected right to stop the system. If a junior employee sees the model acting on a "theft" that didn't happen (as in the Post Office case), can they interrupt execution without fear of punishment? If not, you have no safety valve.

4. Drift Detection

The Rule: Accuracy is not the same as relevance.

The Practice: Models drift because the world changes, not just because code breaks. You need to monitor for semantic decay. Does "Active User" still mean what it meant six months ago? If the meaning has drifted, the system must pause.

The Meaning Mesh doesn't slow innovation.

It prevents catastrophic misinterpretation, the hidden failure mode of AI.

From Concept to Practice: The Meaning Audit

It's easy to agree that meaning matters. It is much harder to operationalize it. When I work with leadership teams, the turning point usually comes when they stop treating "ontology" as an abstract word and start treating it as a daily management discipline.

If you want to protect your organization from semantic drift, start with a Meaning Audit. Before any significant AI deployment, ask these five questions:

1. Do we agree on the meaning of the key terms feeding this model?

   Test: Put your head of product, head of compliance, and lead data scientist in a room. Ask them to write down the definition of "success" for this model. If the papers don't match, stop.

2. Is the tool mathematically fit for the meaning we are assigning it?

   Test: Are we using a probabilistic tool (like a Zestimate) to make deterministic decisions (like buying a house)?

3. Where does human judgment enter the loop, and is that explicit or accidental?

Test: Is there a named role that can say, "The model is statistically confident but contextually wrong"? Or are you relying on the quiet heroism of frontline staff to fix things?

4. What happens when the model is clearly wrong?

Test: Is there a "safe harbor" protocol? Can a junior employee challenge the "system's" decision without fear of punishment? (In the post office, the answer was no.)

5. Who owns the evolution of meaning?

Test: Markets shift. Regulations change. Who is responsible for updating the definitions in the model? If the answer is "no one," you are already drifting.

These questions are not overhead. They are the cheapest insurance policy you will ever buy for your AI program.

We are entering an era of unprecedented volatility. The speed of information is increasing, but the clarity of information is decreasing. AI does not replace human intelligence. It magnifies human interpretation. If that interpretation is grounded in clear meaning, AI allows us to see further and act faster than ever before. But if that interpretation is grounded in ambiguity, bias, or drift, AI will simply accelerate our collision with reality.

Organizations that ignore this will repeat the tragedies of Horizon, Boeing, and Zillow, victims not of model failure, but of meaning failure.

Organizations that embrace it will build systems capable of acting coherently under pressure. They will move from feeling at the mercy of opaque systems to feeling responsible for the meanings those systems encode.

Meaning is not optional. Trust is not emotional. Predictable volatility is the real AI edge.

About the Author

Andrew J. Turner has spent his career at the intersection of scale, systems, and reinvention. In the first half of his journey, he led major transformation programs inside complex global organizations, including GE, SAP, Telefónica, and Tesco, latterly serving as COO of Tesco Mobile. These environments shaped his understanding of how people, processes, and meaning hold together under pressure.

In the second half, Andrew moved into high-growth leadership and advisory roles and is now on his third Silicon Valley startup path. His work focuses on helping technology companies build trust-based systems, strengthen semantic integrity, and scale coherently in the age of AI.

Andrew is the author of the upcoming book Predictable Volatility (2026) and advises founders, investors, and boards on AI readiness and decision risk in complex, high-stakes environments. He hosts The G&T Sessions podcast and holds an MBA from Ashridge.

LinkedIn: https://www.LinkedIn.com/in/iamandrewjturner

CHAPTER 31

THE INDUSTRIAL AI ENGINE: TAMING COMPLEXITY WITH THE DIGITAL TWIN

By: Aasim Waheed
Industrial Translator: Bridging Technology and Profitability
Houston, Texas

> *All models are wrong, but some are useful.*
>
> —George E. P. Box

The AI Edge: Why Industrial AI Demands a Digital Twin Foundation

When the half-a-billion-dollar petrochemical reactor failed at 2:47 a.m. on a Tuesday, the operators had terabytes of data but couldn't answer the CEO's simple question: "Why didn't we see this coming?" The data existed. The sensors worked. The problem? The data lived in 17 different systems that couldn't talk to each other, and no one knew which "temperature reading" actually mattered.

The promise of AI for industry is immense: optimizing operations, predicting failures, and revolutionizing efficiency. And yet, for many industrial leaders, this vision collides with a harsh reality: their physical assets, legacy systems, and chaotic data create a "swamp" that paralyzes AI's true potential. You cannot build a sophisticated AI "penthouse" on such an unstable foundation. To truly unlock the "AI edge" and thrive in this disruption, industrial enterprises must first construct a robust, real-time "world model" of their physical assets. This foundational platform, the digital twin (DT), is the indispensable engine that tames complexity and transforms raw industrial data into AI-ready intelligence, making the AI future a tangible reality.

Read on to discover how.

Demystifying the Digital Twin: An Analogy

The term "digital twin" (DT) is buried in jargon. To understand it, let's discard the technical definitions and use a relatable analogy: building a custom house.

First, you work with an architect to create blueprints, 3D models, and specifications. This is the *design twin*. It's the "as-designed" state, a virtual representation of what the house *should* be, existing before a single shovelful of dirt is moved.

Next, construction begins, and reality immediately sets in. The plumbing contractor uses a different pipe vendor due to availability. The electrician runs wiring through a different path. The builder hands you a set of "as-built" drawings, marked up in red ink, that show what was *actually* built. This is the *as-built twin*, the "as-is" state of your new home.

You move in, and the house becomes a living asset. Smart thermostats report temperature, security cameras show who is at the door, and smart meters track energy use. This stream of live, operational data is the *real-time twin*. It's the "what is happening *now*" state.

Finally, six months later, the air conditioner breaks. A technician logs the repair, noting the parts, date, and cost. This service history, along with all your warranty documents and manuals, forms the *maintenance twin*. It's the "what has *happened*" state.

A true digital twin is not any one of these things. It is the seamless and continuous integration of all four. It is a single, trusted platform where you can see the *as-built* 3D model of your AC unit, view its *real-time* temperature, and instantly pull up its full *maintenance* history and original *design* specifications. This integrated, contextualized whole is the foundation of digital intelligence.

The Foundational Challenge: Bridging Three Worlds

Applying this "house" concept to a sprawling, high-risk industrial facility, whether a pharmaceutical production line, an automotive factory, or an oil refinery, magnifies the complexity exponentially. The four layers of the twin are not built by one team; they are sourced from three distinct, and often conflicting, technology domains, each with its own culture, priorities, and data.

1. Operational Technology (OT): This is the world of the physical asset. It includes the sensors, control systems (DCS, distributed control systems; PLC, programmable logic controllers), and historians that run the plant. OT's primary mandate is *safety and availability*. Its culture is risk-averse, and its technology is built to last for decades. This domain provides the data for the *real-time twin*.

2. Engineering Technology (ET): This is the world of design and physics. It includes the 3D models, CAD drawings, and simulation software that define the asset. ET's mandate is *accuracy and integrity*. This domain provides the data for the *design twin* and the *as-built twin*.

3. Information Technology (IT): This is the world of business systems. It includes the cloud platforms, ERP (enterprise resource planning) systems, maintenance logs

(enterprise asset management/computerized maintenance management system, EAM/CMMS), and data analytics tools. IT's mandate is *data confidentiality and agility*. This domain provides the data for the *maintenance twin* and the infrastructure to host the entire platform.

The central challenge of building a digital twin is not technical; it's organizational. It is the challenge of bridging the "clash of cultures" between these three domains to create a single, unified data flow.

The Data Liberation Challenge and the "Data Refinery"

Before you can even begin integration, you face a more fundamental problem: Your data is being held hostage. For decades, industrial assets have been built with proprietary, closed systems. The data from Vendor A's control system often cannot be easily read by Vendor B's analytics platform. Every attempt to connect systems requires a costly, brittle, custom-built interface. This is the "integration tax" that has silently crippled industrial innovation.

Furthermore, the raw data that *is* accessible is like "digital crude." It's abundant but unusable in its native state. A sensor tag named TI-10-47b is meaningless without context. Does it measure in Celsius or Fahrenheit? Is it on a critical compressor or a non-essential pipe? Is it currently in service or offline for maintenance?

This is where AI hits the wall. You cannot feed "digital crude" to an AI model and expect intelligent results. You must first liberate and refine it. This process is the "data refinery." It is the hidden engine of your digital twin.

1. Liberation: This is a procurement and strategy fight. It means reclaiming ownership of your data by demanding open, non-proprietary standards from all new vendors.

2. Transportation: This is a secure IT/OT architecture. It means building secure data pipelines that can move high-volume operational data from the plant floor (OT) to the

enterprise cloud (IT) without compromising the safety of the facility.

3. Refining: This is the most critical step. It is the process of contextualization. This is where you fuse the different data streams together. The raw sensor tag TI-10-47b (*real-time twin*) is automatically linked to its 3D model (*as-built twin*), its engineering specs (*design twin*), and its full repair history (*maintenance twin*).

Only after this "refining" process do you have a high-grade, AI-ready fuel. The raw tag TI-10-47b has been transformed into a trusted, intelligent object: "Reactor 1, Inlet Temperature, Post-Preheater, Last Serviced: 6-Oct-2025." Now, an AI model can finally understand it, and a human can finally trust it.

The Human Operating System: A Strategy for People

Building this platform is a massive technical undertaking, but it is a socio-technical one. Technology is only half the battle. The other half is your people. As a leader, you must address the human element with the same rigor you apply to the technical architecture.

This begins by addressing the natural fear of change. For a senior operator who has trusted their "gut feel" for 30 years, an AI-powered recommendation can feel like a threat. The narrative must be clear: This technology is not here to replace their expertise; it's here to give it superpowers. It augments their intuition with data, allowing them to see a failure coming weeks in advance rather than just minutes.

To manage this new asset, you need a new organizational structure. The most effective model is a center of excellence (CoE), a cross-functional team co-owned by technology and operations leadership. This CoE is the engine for your digital twin, acting as the hub for builders, data guardians, frontline optimizers, and strategists. Critically, this CoE must work with HR to build new career paths. A talented process engineer who shows an aptitude for data should see a clear path to becoming a "digital twin analyst" and, eventually, a "senior operations optimizer." This transforms the "threat" of AI into a tangible, exciting career opportunity.

The Scaling Imperative: From Pilot to Enterprise Value

The journey does not end with one successful twin. The real value, the kind that moves a stock price, comes from scaling. This is the COO's primary challenge: navigating the "pilot trap." A pilot project often succeeds in a protected bubble. Scaling it across a global enterprise is where most initiatives fail. The solution is a disciplined, phased approach: crawl, walk, run.

- Crawl (Replicate): Start with one pilot on a single, high-value asset. Prove the technology and, critically, the financial ROI (return on investment). Use metrics like overall equipment effectiveness (OEE), reduced maintenance costs, and improved safety incidents. This is the business case you take to the CFO.

- Walk (System-Wide Integration): Replicate the success on three to five similar assets. Then, expand the twin to cover an entire process unit (e.g., a distillation column and its associated equipment in a chemical plant or an entire assembly line in a factory) to unlock system-level optimization.

- Run (Enterprise Scale): Connect multiple unit-level twins to create a facility-wide or even value-chain-wide view. This is where you move from optimizing a single asset to optimizing the entire business.

This "Crawl, Walk, Run" model is the ultimate use case. It proves that your value is repeatable, scalable, and financially sustainable.

The "Day 2" Problem: Defending Your Twin from Data Entropy

You've done it. You've scaled your twin. The original project team moves on. Now, the final, permanent battle begins: the fight against data entropy. Data entropy is the natural tendency for a clean system to decay into chaos. A new vendor installs a pump with a non-standard tag. A system upgrade breaks a data pipeline. A sensor's calibration drifts. Slowly, the "data swamp" creeps back in. The twin's data becomes untrusted, its AI models become inaccurate, and your multi-million-dollar asset regresses into an expensive, unused dashboard.

A digital twin is not a project to be completed. It is a living asset to be managed. This requires a permanent two-pronged defense:

1. Organizational Defense (The Governance Council): The CoE evolves into a permanent governance body. It is staffed with "data owners" (the business leaders accountable for value), "data stewards" (the frontline guardians responsible for quality), and "data custodians" (the IT/OT teams who manage the infrastructure).

2. Technical Defense (The Automated Pipeline): The "data refinery" is upgraded with an immune system. This automated data pipeline validates all incoming data against your standards. Any non-compliant data (e.g., a "temp_test" tag) is automatically quarantined and an alert is sent to the data steward. It enforces the rules, so your people don't have to.

Common Pitfalls and How to Avoid Them

The digital twin journey is filled with a lot of traps; many are predictable. Here are the most common pitfalls and how to address them.

Pitfall 1: Treating the Twin as an IT Project

The most common mistake is to treat the digital twin as a piece of software, another dashboard for the control room. It inevitably fails. The solution is to treat it as a business transformation initiative, led by operations. The goal is not to see the data; the goal is to change how decisions are made, how operators respond, and how maintenance is scheduled.

Pitfall 2: Starting with the Wrong Use Case

Many teams get this backward. They either pick a low-value, simple asset that impresses no one, or they pick an incredibly complex, high-value asset where the physics are poorly understood. The first fails to get future funding; the second fails to deliver results. The key is to pick a high-impact, well-understood asset or process with clear, measurable KPIs (key performance indicators). Start with a "win" that is both meaningful and achievable.

Pitfall 3: Underestimating the Data Foundation

You will always hear, "Let's do the fancy AI model first," and skip the "boring" data plumbing. That is a fatal error. As I've noted, the "data refinery" is the essential enabler. You must have a strategy to liberate, transport, and refine your data before you can build anything of value on top of it.

Pitfall 4: Ignoring People and Change Management

This is the silent killer. A perfect digital twin that no one trusts or uses is worthless. Adopting this technology means creating new processes, new roles, and new skills. You must address the human element from day one. This includes building the governance, creating the new career paths, and managing the cultural change with the same rigor as the technical build.

Pitfall 5: Overlooking AI and Integration Possibilities

If your twin is just a replica, a 3D model with live data, you are leaving 90% of the value on the table. The true power comes from embedding analytics and AI into the twin. It should be a platform for prediction (e.g., predictive maintenance), simulation (e.g., "what-if" scenarios), and optimization (e.g., AI-driven process control).

The Way Ahead: A Leader's Checklist

If you are a CIO, CTO, VP of operations, or senior decision-maker, here is a checklist to ensure you capture the "AI edge" through your digital twin initiative.

- Strategic Alignment: Confirm how the digital twin initiative aligns with your enterprise's core strategic priorities (e.g., asset reliability, cost optimization, new service revenue, sustainability).

- Value Hypotheses: Articulate clear, measurable value hypotheses. For example: "We believe a digital twin will reduce unplanned downtime by 15% on our critical assets" or "This will enable a new twin-enabled service for our customers."

- Governance and Sponsorship: Secure senior executive sponsorship (ideally a COO or business-line VP) and form a cross-functional steering committee (IT, OT, ET, Operations, Finance) from day one.

- Data Foundation Audit: Conduct a rapid audit of your data landscape. Do you have the sensors? Can you access the data in real time? Are your data models standardized? Where are the quality gaps?

- Pilot Selection (The "Crawl"): Choose your first asset or process wisely. It must be high impact, but also have manageable complexity and measurable metrics.

- Model Strategy: Develop your modeling approach. Will it be physics-based (using engineering principles), AI-based (using machine learning on historical data), or a hybrid? This depends on your asset and organizational maturity.

- Scalability Plan (The "Walk and Run"): Design for scale from the beginning. Your architecture, governance model, and data standards must be repeatable, or you will be trapped in the pilot phase forever.

- Business Model Innovation: Do not restrict the twin to internal optimization. Explore external services, digital-asset business models, and outcome-based contracts with your customers.

- Capability Build-Out: Develop in-house skills and strategic partnerships. A digital twin requires a blend of data science, modeling, simulation, domain knowledge, and change management.

- Outcome Tracking and Communication: Define your success metrics, track progress relentlessly, and communicate wins, especially early wins. A digital twin that goes "quiet" will lose momentum and funding.

The Digital Twin as Your AI Edge

In the industrial world, AI cannot thrive on its own. It needs a model of the physical world that is trusted, real-time, and rich with context. A digital twin is not just another piece of software. It is a profound, strategic, and socio-technical system. It is the organizational and technical framework for taming industrial complexity. It is the "data refinery" that turns digital crude into AI fuel. It is the bridge that connects the three disparate worlds of IT, OT, and ET. And it is the human operating system that empowers your people with new skills and new careers.

The "AI edge" for our civilization's most critical industries will not be found in a standalone algorithm. It will be found in the deep, foundational work of building this platform. The companies that thrive in the next disruption will be those who understood that the digital twin was never the end goal. It was the essential, non-negotiable beginning.

About the Author

Aasim Waheed is a distinguished expert in industrial digital transformation with over two decades of experience at the intersection of heavy industry operations, control systems, and data technology. He specializes in bridging the critical gap between operational technology (OT) and information technology (IT), architecting complex data strategies to drive business value. A leading voice in IT-OT-ET integration and digital twin implementation, he advises senior executives on navigating the technical, financial, and cultural challenges of building the intelligent, AI-ready industrial assets of the future. He holds masters in data analytics from Georgia Tech, and has completed various programs at MIT, Chicago Booth, and UT Austin. He has led major digital initiatives for some of the world's largest industrial corporations.

Email: aasimwaheed1@gmail.com

LinkedIn: https://www.linkedin.com/in/aasimwaheed/

CHAPTER 32

AI IS NOT A DOMINO PIECE: WHY IMPLEMENTING IT IS A PUZZLE, NOT A CHAIN REACTION

By Daniel Wielander
AI Compiler for Corporates
Vienna, Austria

What truly sets the "new age" of AI apart, especially since the rise of OpenAI and ChatGPT, is the incredible speed at which new possibilities, features, and tools are emerging. AI is no longer evolving in a slow, step-by-step manner; it's accelerating exponentially. Because almost anyone can now interact with AI using natural language, it's no longer just a topic for experts. It's becoming relevant for every employee in a company.

Over the past seven years of my consulting work, I've noticed recurring patterns across many organizations. Unfortunately, these patterns don't necessarily lead to success. Here's my personal "best of" list of common, but not always helpful, approaches I've seen, including some ideas on how to solve them.

Without a Route, Every Peak Is Just a Mirage

In both the public and private sectors, I've observed two main approaches to adopting AI. A small number of organizations begin with strategic considerations, asking where AI can truly add value, how it should be used (for example, the desired level of autonomy), where it should not be applied, and what challenges it might bring.

However, the majority of organizations jump straight into experimentation, launching pilot projects or so-called proof of concepts (PoCs) to quickly test the technology. At this stage, there are usually no clear guidelines, just a problem that AI is expected to solve. While both approaches have their merits, the strategic route is often overlooked because it requires upfront investment in frameworks without immediate visible results.

The second approach, though more common, comes with risks. Most PoCs lack clearly defined, measurable goals, such as what exactly should be achieved, how success will be measured, and when the solution should move into production. As a result, many of these projects remain stuck in a perpetual test phase, never delivering real operational value. I often refer to this stage as the "AI graveyard."

Another issue is the frequent misuse of the term "proof of concept." Originally, it meant testing whether a specific AI method could solve a particular problem. But for 90% of organizations, there are already countless proven examples, so another test isn't needed. What's needed is actual implementation in the production environment.

Advice: To unlock AI's full potential, organizations need both a strategic framework with clear goals and boundaries and the execution of use cases that deliver measurable value.

Use Case Hunters

About six years ago, I often had to work hard to convince C-level executives that AI could offer solid solutions to specific business problems. Fast forward to today, and the rise of generative AI has flipped the narrative. AI is now seen as a "silver bullet" capable of solving virtually anything. For many users who had no access to AI tools before, it feels like they've discovered the Holy Grail. Meanwhile, for IT leaders in the same organizations, it can feel more like Pandora's box has been opened.

AI should never be used for its own sake. It's a powerful tool but just one of many in the toolbox. And with great power comes great responsibility. Before jumping to AI as the solution, it's worth asking some fundamental questions: Do we even need this process? Are all the steps necessary (such as approvals)? Could the process be redesigned entirely?

Only after optimizing the process should AI be considered as a potential solution. A classic example is route optimization. Imagine a delivery company with over 30 employees manually planning delivery routes based on their experience. By applying AI to this existing process, efficiency could likely be improved by at least 30%. But if we rethink the process entirely, say, by using drones for delivery where legally allowed, the efficiency gains could be even greater, thanks to shorter delivery paths.

Another common challenge I've seen, especially at the beginning of an organization's AI journey, is the use of the wrong criteria for selecting initial use cases. Many companies aim for the use cases with the highest return on investment (ROI). While this makes sense financially, it often overlooks a key point: The complexity of implementation usually increases with the potential ROI. So, while the financial upside may be high, the difficulty of execution is equally high.

It's like asking a newly graduated doctor to perform one of the most complex surgeries, separating conjoined twins joined at the head. It's simply not the right starting point.

Early AI use cases should be representative of the broader organization yet as simple as possible. This allows the organization to learn and build confidence before tackling more complex challenges.

Advice: Start small and simple to build momentum, rather than diving into the hardest-highest ROI project first.

Center of AI Death (CoAID)

I'm always amazed at how quickly organizations find ways to develop AI use cases at the beginning of their journey. Often, this happens through internal IT teams or external service providers. Over the years, I've seen two common approaches emerge.

The first involves organizations that simply continue implementing one use case after another without giving much thought to governance or long-term operations. Like any software solution, AI systems need to be maintained and supported. The challenge with learning AI systems is that they can evolve in unintended ways over time, a phenomenon known as "model drift." Initially, existing IT teams might be able to handle support, but as more use cases go live, it becomes increasingly difficult to maintain them reactively, let alone proactively. This lack of foresight can lead to serious issues, including a loss of trust among employees. I've seen several AI initiatives fail at this critical point.

The second approach is more strategic: embedding AI within the organization. This often involves creating a dedicated unit within IT or establishing a new function, commonly called an "AI Center of Excellence" (CoE). This model focuses on building centralized expertise and offers end-to-end services, from identifying use cases to developing and operating AI solutions. I support this approach, especially in the early stages, provided it's integrated into existing organizational structures.

What do I mean by that? Many companies set up separate CoEs for digitalization, process automation, and AI. The problem is that each CoE tends to push its own agenda. Experts within each unit naturally try to solve problems using their specific tools and technologies. What's often missing is an objective, cross-functional perspective, someone who can evaluate a problem neutrally and determine the best approach, regardless of the technology.

That's why I advocate for a unified CoE focused on organizational improvement as a whole. Its sole purpose should be to make the organization more successful, with technology selection being a secondary concern.

A well-structured CoE lays the foundation for a sustainable AI journey. It defines governance, assigns responsibilities, and establishes clear processes for every phase of the AI lifecycle. However, as organizations mature, I've noticed two common pitfalls. First, while every CoE aims to enable scalability, this only works for use cases, not necessarily for human resources. As the number of use cases grows, the CoE must also scale up its team. This can become costly and challenging, especially during economically difficult times.

Second, there's often a gap in domain-specific process knowledge. Business units understand their processes and data far better than AI experts do. On the other hand, AI specialists understand the models but can't be expected to master the nuances of every business process. This disconnect can lead to inefficiencies in further development or troubleshooting.

The solution, as is often the case, lies in the middle. In process automation, this is known as the "citizen developer" model. Initially, the CoE leads development, but over time, responsibility gradually shifts to the business units. These teams begin to build their own AI capabilities and even hire AI talent. The CoE then transitions into a quality assurance role, develops central AI solutions for the entire organization, evolves governance frameworks, and monitors market trends.

Advice: Establish a central automation team (CoE) early on, and start empowering business units to take ownership, combining centralized oversight with domain expertise.

The Most Underestimated Competence

One of the most fundamental shifts since the emergence of ChatGPT and similar tools is how people perceive and define artificial intelligence. Before the era of publicly accessible generative AI, only a small portion of society had a clear understanding of what AI actually meant. Today, nearly everyone has formed some kind of opinion or expectation about AI, whether accurate or not.

The challenge is that these perceptions vary widely. Few technologies have sparked such a diverse range of interpretations regarding what they are, what they can do, and what risks they pose. I believe this is due to two main factors: First, there is still no globally accepted definition of artificial intelligence. Second, the hype surrounding generative AI has suddenly given almost everyone the opportunity to interact with AI tools firsthand. The result can be like building a high-tech car and then expecting elementary school children to drive it.

The widespread availability of AI should primarily be seen as an opportunity rather than a risk. However, there are some downsides. One major issue is that employees often lack a clear understanding of the technology. This can lead to unrealistic expectations on one hand and fears of job loss due to automation on the other.

I frequently encounter a recurring situation when it comes to the use of generative AI: Companies introduce a solution based on generative AI and are then surprised when it's not widely adopted, or worse, when it receives a lot of negative feedback. Here are a few typical comments I've heard from within organizations:

- "The solution hallucinates. It gives incorrect answers that aren't even in the data."

- "I know the document exists, but the AI doesn't use it."

- "The AI only provides very generic responses that don't help me at all."

When things go wrong, AI is usually the first to be blamed. Of course, every AI solution makes mistakes. That's a given. But when we take a closer look, we often find that over 70% of the issues actually stem from user errors. These mistakes range from misunderstandings about how the solution works (for example, confusion between data connectivity, language models, and knowledge models) to a lack of basic prompting skills.

In the tech world, there's a huge number of specialized terms that are often misused, usually unintentionally. That's why it's essential to help employees understand the key concepts in the AI space: what the terms mean, what types of AI solutions exist, and what they're best used for.

Equally important is building user competence when it comes to working with AI. This means helping people understand their role in using AI effectively. On one hand, it's about learning how to operate AI tools, especially how to "prompt" generative AI systems. On the other hand, it's about recognizing the responsibility users have to critically evaluate the results AI provides.

This responsibility applies not only to generative AI but to all types of AI solutions. In my experience, effective prompting is by far the most underestimated skill, when it comes to successful AI adoption. Many people still think prompting just means typing a few words, like they would in a Google search, and waiting for an answer. But it's much more than that.

Advice: User understanding and skills are often the weakest link. Emphasizing education in AI basics and training staff to craft good prompts and think critically is key.

Culture Eats AI for Breakfast

Organizations that are truly committed to using AI have long understood that the most important piece of the puzzle is company culture. But changing culture doesn't happen at the push of a button or through a short-term project. Unfortunately, this is exactly how many companies still approach it.

Just like when the internet was first introduced, it takes time and deliberate initiatives to help employees understand what AI can do, how their roles (and those of others) might evolve, and how these changes can positively impact their own work. Clear communication and a shared vision are essential.

Equally important is giving employees the chance to quickly gain hands-on experience with AI. Early exposure helps build confidence and understanding. This also ties into the concept of a healthy error culture, something that is still underdeveloped in many parts of Europe. The ideal environment for automation and AI is one where mistakes are seen as learning opportunities that drive growth. After all, AI thrives on experimentation and failure, and a culture of fear suffocates this process.

If cultural change is ignored, the consequences can be felt quickly. People are creatures of habit. It's a basic instinct that gives us a sense of security. Any disruption to the familiar can lead to uncertainty and even fear. If these concerns aren't addressed, a strong "anti-AI" sentiment can emerge, marked by resistance and rejection of the technology.

Advice: Culture—encouraging exploration, addressing fears, and normalizing learning from mistakes—is the make-or-break factor for AI success.

In summary, implementing AI is not a simple domino effect where one change automatically leads to the next; it's a multifaceted puzzle. Success requires aligning strategy, process selection, organizational structures, employee skills, and company culture. By addressing all of these pieces together, instead of expecting a chain reaction from a single AI initiative, organizations can truly unlock AI's full potential and avoid the pitfalls of the "AI graveyard."

About the Author

Daniel Wielander is a management consultant specializing in process automation and artificial intelligence at Ernst & Young Austria. For the past five years, he has led the AI-enabled automation division, which focuses on guiding organizations from a strategic perspective in the adoption and scaling of process automation technologies. Daniel's core expertise lies in advising public sector institutions, helping them identify the right building blocks to ensure their automation initiatives deliver measurable value. In addition to his consulting work, Daniel is a keynote speaker and the program director of an AI course with a focus on finance and controlling.

LinkedIn: https://www.linkedin.com/in/daniel-wielander/

FROM AUTOMATION TO AUTONOMY: THE EXECUTION DISCIPLINE THAT TURNS AI ADOPTION INTO ADVANTAGE

By Irene Yang
Product Innovation & AI Transformation Leader
New York, New York

The Transformation Imperative

Artificial intelligence has emerged as one of the most disruptive forces to hit business and society since the internet. Its latest inflection point, being driven by generative AI and increasingly agentic systems, is moving AI into everyday products and operations. It's reshaping how organizations compete, how work gets done, and how professionals create value across nearly every field. But the noise-to-signal ratio is high: The same headlines that celebrate breakthrough models also fuel anxiety about displacement, disruption, and whether leaders can keep up.

The story of this decade won't be who adopted AI first. It will be who turned adoption into durable advantage. Many organizations have launched pilots and copilots. Fewer have redesigned end-to-end workflows, strengthened data foundations, and built the governance, evaluation, and change muscle needed to scale responsibly. This chapter separates signal from noise: It maps the shift from automation to autonomy, highlights the AI trends most likely to matter, and translates them into practical guidance for leaders and professionals.

The AI Landscape: From Automation to Autonomy

AI is not new. Traditional machine learning, predictive analytics, optimization, computer vision, robotics, and rule-based systems have powered enterprise automation for decades, from fraud detection and algorithmic trading to supply-chain forecasting and smart manufacturing. What is new is accessibility. When OpenAI launched its free version of ChatGPT in late 2022, it went viral gaining over a million users in a few days. The AI landscape shifted not because the technology became more ambitious, but because it became more usable. Large language models (LLMs) made natural language a practical, universal interface for mainstream adoption.

Generative AI (GenAI) differs from prior systems in two important ways. First, it can produce new content, text, code, images, audio, video, by learning statistical patterns from massive datasets and generating context-appropriate outputs. Second, it can generalize across tasks. This broad competence explains GenAI's rapid evolution from novelty to infrastructure: It lowers the cost of experimentation, shortens time to prototype, and expands who can build useful tools.

Yet these capabilities, while transformative, represent just one stage in a longer evolution. GenAI is one point on a broader arc of automation to autonomy. The next phase extends beyond better chatbots to systems that can plan multi-step workflows, act through enterprise tools, coordinate across functions, and verify results – all while maintaining safety, accountability, and auditability.

A useful mental model is a spectrum of AI agency. Agency can be defined as the ability to pursue goals with partial autonomy: to interpret context, plan steps, call tools, maintain state, and adapt when conditions change. This spectrum ranges from human-led assistants to constrained agents (operating within defined policies) to agentic systems that orchestrate multiple specialized components across workflows. In practice, most organizations will operate across all three modes:

- Automation: Rules and ML models that execute predefined decisions

- Assistance: Copilots that draft, summarize, and recommend while humans remain primary operators

- Autonomy: Agentic systems that plan, act through tools, maintain state, and escalate exceptions under clear policies

The strategic question is not "Should we automate?". It's where automation is appropriate, where must it remain advisory, and what controls make it safe.

Market Reality and Adoption

Market forecasts vary widely (and should be treated as directional, not deterministic), but the underlying trajectory is clear: the AI market is experiencing exponential growth, driven by rapid advancements in computational power, massive capital investment, and surging adoption across various industries. AI spend is compounding rapidly, and GenAI is taking a larger share of that spend. Precedence Research estimates the global AI market revenue at roughly $638B in 2025 with forecasts rising into the multi-trillion range over the next decade.[1] Its generative AI estimates are roughly $48B in 2025 with a substantially higher CAGR than the broader AI market.[2] These numbers are useful mainly because they indicate the size of the prize and why every major platform vendor is racing to industrialize AI.

Enterprise behavior reflects this trajectory. McKinsey's 2025 global survey reports that 88% of respondents say their organizations use AI in at least one business function.[3] In the US, Gallup surveys find workplace AI use have nearly doubled from 21% in 2023 to 40% in 2025, with frequent use rising from 11% to 19%.[4] Employee usage is rising sharply as tools become embedded in daily workflows. [4] OpenAI and Wharton reports also identify technology, banking/finance/insurance, professional services, manufacturing and healthcare and life sciences as the fastest-growing sectors and those operating at the largest scale.[5,6]

The Value Realization Challenge

However, adoption is not the same as value. Throughout 2024-2025, the most common enterprise failure mode has been deploying AI as an add-on or in piecemeal, a pilot here, a chatbot there, a coding assistant for a few teams, without truly redesigning workflows, upgrading data foundations, or changing accountability. The result is a paradox many leaders recognize: widespread AI experimentation with limited measurable bottom-line impact.

Successful firms pulling ahead treat AI as a holistic, strategic transformation program, not tool deployment. They invest disproportionately in data quality, integration, process redesign, and change management. They also make value measurable from the start, not as vague productivity lifts, but as cycle-time reduction, error-rate improvement, revenue lift, cost-to-serve reduction, and risk reduction, all with clear baselines and owners. They embrace best practices and discipline in vigorously defining and tracking leading and lagging metrics alongside learning goals.

The debate has long shifted from "Is AI real?" to "What does it take to scale it safely, economically, and with measurable value?" Competitive advantage in the age of AI will come from disciplined execution at scale.

Top Technology Trends Reshaping Enterprise AI

Trend 1: GenAI Moves from Content to Capability

GenAI's first wave focused on drafting, summarizing, and answering questions. The second wave shifts from content to capability given model advancements such as improved reasoning, API calling, executing code, and workflow orchestration. Models are being embedded inside business systems with tighter guardrails, better grounding in enterprise data, and measurable impact on end-to-end outcomes.

A practical litmus test is simple: If a use case cannot be instrumented (inputs, outputs, quality, and business impact), it will not scale. Winners will operationalize GenAI like any other mission-critical system: monitored, tuned, secured, and continuously improved.

Tip: Leverage GenAI's new capabilities. Redesign end-to-end workflows and embed GenAI within business systems. Measure its success by whether it changes the economics of a workflow.

Trend 2: RAG and GraphRAG Ground Models in Enterprise Truth

LLMs generate from patterns in training data; without grounding, they can be outdated or wrong. Retrieval-augmented generation (RAG) reduces this risk by retrieving relevant enterprise context at query time and using it as evidence for responses.

Traditional RAG works well for finding and summarizing tasks but can struggle when questions require connecting entities and relationships across documents, policies, transactions, and time. GraphRAG addresses this by building a knowledge graph during ingestion and using that structure at query time to support multi-hop reasoning and more reliable summarization. [7,8] In regulated industries, the "why" matters as much as the "what." Knowledge graphs support provenance, traceability, and transparent evidence chains.

Takeaway: Enterprises will not win by prompting better; they will win by grounding models in trusted data with traceability.

Trend 3: Agentic AI Shifts from Assistants to Controlled Autonomy

Agentic AI represents the next leap beyond generative copilots. While copilots accelerate human performance by assisting with tasks, agentic systems go further – they take on portions of the work themselves. These systems interpret context, plan steps, call tools, maintain state, and adapt when conditions change. According to Gartner, by 2028 agentic AI will autonomously handle 15% of daily business decisions and be embedded in one-third of enterprise software.[10]

The value proposition is strongest in high-volume, rules-driven processes – yet so are the risks, which is why Gartner predicts many early initiatives will stall without disciplined governance, cost and value management.[9,10]

The implementation principle is simple: Build for disciplined autonomy. Clearly define what an agent can do, what it cannot do, how it is monitored, and how it escalates.

Takeaway: Agentic systems will scale only when autonomy operates within explicit boundaries of policy, permission, and oversight – effectively functioning as a new class of digital workers.

Trend 4: Small Language Models (SLMs) Become Strategic Workhorses

Large models are powerful but can be expensive to run continuously. Small language models (often under ~10B parameters) offer lower cost, lower latency, and easier deployment on premises or at the edge. The trend is moving towards domain-specific models that can run effectively on smaller, tuned models paired with strong retrieval and guardrails. What enterprises need most is reliable performance in narrow contexts – precisely what SLMs deliver at scale.[11]

SLMs help unlock strategic advantage. SLMs enable a routing strategy that is multi-model: route tasks to the smallest model that meets quality requirements and escalate to larger models only when needed. Many enterprise tasks and use cases may rely more on domain-specific knowledge and contextual insights versus requiring the vast data and computation in larger models.

Takeaway: Design an "AI portfolio" that routes tasks by cost, latency, and risk, not a single-model strategy.

Trend 5: Agentic RAG Enables Dynamic Knowledge Access

RAG retrieves information, agentic RAG reasons about retrieval. Instead of a single query-and-answer pass, an agent can iteratively refine queries, reconcile conflicting sources, and verify evidence across multiple tool calls.[12] This becomes critical for multi-step processes such as claims processing, procurement, legal review, incident response, and customer issue resolution where "getting the right answer" requires structured verification.

Takeaway: As workflows become multi-step, retrieval must become task-aware, auditable, and iterative.

Trend 6: Multi-Agent Orchestration Mirrors How Organizations Work

Just as retrieval must become task-aware, complex work itself requires coordination across specialized agents. Complex work rarely happens in a single brain. It happens through coordination: specialization, handoffs, checks, and shared goals. Multi-agent systems apply the same logic: A coordinating orchestrator routes tasks to domain agents and integrates results.

At enterprise scale, orchestration forces hard questions about identity, permissions, evaluation, logging, and escalation designed upfront so failures are visible, containable, and correctable.

Takeaway: If you cannot observe an agent's actions end to end, you cannot safely scale it.

Trend 7: Open Standards and Open Source Become the Integration Layer

Enterprises are adopting a multi-model reality: mingling proprietary models with open-source alternatives and mixing vendor platforms with internal tooling. Open standards matter because they reduce integration friction, broaden talent pools, and limit lock-in.

Model context protocol (MCP), developed by Anthropic and donated to the Linux Foundation in December 2025 as a founding project of the Agentic AI Foundation, is one example: a standard for connecting models and agents to tools and data sources in a consistent way.[13,14]

Takeaway: Integration and governance will decide the winners; standards reduce the cost of both.

Trend 8: Partner Ecosystems and Marketplaces Become a Competitive Surface

The enterprise AI stack has expanded beyond "cloud + software + services" to include eight categories: chipmakers, model providers, orchestration platforms, vector databases, data vendors, governance tools, and industry-specific agents. No single vendor can supply all of it well, and it's clear that partnerships and ecosystem strategies need to be integrated into solution architecture. As a result, major providers are placing bets on marketplaces. Microsoft's marketplace launch highlighted thousands of AI apps and agents integrated into products such as Azure AI Foundry and Microsoft 365 Copilot,[15] while AWS announced new innovations for building AI agents, expanded marketplace listings, and made a $100M investment to boost agentic AI development.[16]

This marketplace expansion creates both opportunity and risk. CIOs face tremendous pressure to deliver AI value while managing the risks of architectural sprawl. This sprawl contributes to technical debt and complexity, operational inefficiencies, security risks, and increasing costs, all of which can hinder innovation.

Takeaway: Ecosystem strategy should be part of core architecture design: Build flexible infrastructure, design for modularity, manage partner performance and risk, and protect IP boundaries while still moving fast.

Trend 9: Synthetic Data Expands What Can Be Modeled

Synthetic data can amplify rare edge cases, reduce privacy barriers, and speed development when real data is sensitive or incomplete. It can also improve evaluation by generating controlled test sets that stress models under known conditions.

But synthetic data is not a free pass. If the synthetic distribution diverges from reality, it can import bias or create false confidence. Used well, it is a real force multiplier; used poorly, it quietly breaks reliability at scale.

Takeaway: Treat synthetic data as a controlled instrument: Validate it against real outcomes and use it to strengthen testing, not to avoid it.

Trend 10: AI Assurance and Evaluation Operations Become Mandatory for Scale

As AI systems move from copilots to agents, the bottleneck shifts from "Can we build it?" to "Can we trust it in production?" The next frontier is assurance: continuous evaluation, monitoring, and controlled change management for models and agentic workflows.

Evaluation operations (EvalsOps) is an emerging discipline in AI focused on systematic evaluation, automated testing and benchmarking, continuous monitoring and improvements, observability, and governance to ensure regulatory compliance and quality control gates. Given that LLM models can be non-deterministic with unpredictable outputs, there need to be systematic, effective ways to ensure alignment, trust, and adaptability throughout cycles with continuous stakeholder feedback. This includes testing critical tasks, "red-teaming," which simulate adversary cyberattack, regression testing when prompts or models change, quality metrics tied to business outcomes, and audit logs that allow reconstruction when something goes wrong.

Takeaway: EvalsOps is to AI what DevOps was to software: the operating discipline that turns prototypes into production systems.

Navigating the AI Transformation Journey

The shift to AI-driven operations is an operating-model change as much as a technology change. Most organizations do not fail because the model is not smart enough. They fail because ownership is unclear, data access is weak, workflows are unchanged, and value is not measured.

For Enterprise Leaders: Six Strategic Imperatives

1. **Treat AI as business transformation.** Set ambition, risk appetite, decision rights, and a portfolio view across functions.

2. **Make data readiness a first-class program.** Prioritize access, quality, lineage, and "golden sources" for critical processes.

3. **Build a value dashboard early.** Define baselines, owners, and stop/go gates before scaling.

4. **Redesign workflows end to end.** Clarify where humans approve, where AI advises, and where automation is

permitted. Then update roles, controls, and training accordingly.

5. **Operationalize responsible AI.** Embed privacy, security, documentation, and evaluation into the build-and-run process.

6. **Build ecosystem flexibility without losing control.** Use standards, enforce identity and permissions, and continuously manage vendor risk.

IBM's 2025 CDO study highlights the gap between AI ambition and data readiness, and PwC's 2025 Responsible AI survey links responsible practices to performance, but both imply the same operational point: Governance only works when it is engineered into workflows.[17,18]

For Individuals: Three Pathways to Thrive

Organizations are moving from traditional hierarchies to agile, skills-based ecosystems where human-AI collaboration is essential for competitive advantage. So what should individuals do to thrive in this workforce of the future? First, develop new skill profiles. Three patterns are emerging in high-performing organizations: (a) **M-shaped supervisors** are broad generalists fluent in AI who can orchestrate across multiple domains and successfully manage workflows by providing interdisciplinary and integration leadership (e.g. project leaders who can supervise teams of AI agents handling end-to-end workflows like customer onboarding or financial closing); (b) **T-shaped specialists** blend domain mastery with AI fluency, broad collaboration literacy and execution (e.g. compliance experts who use AI for routine reviews and can handle exceptions, quality, and risk); and (c) **AI-augmented frontline workers** focus on human interaction and judgment while offloading routine tasks (e.g. customer service or account reps who use AI for research and documentation but personally manage more sensitive conversations).

Second, overcome the relevance barrier. Many professionals don't adopt because they don't believe AI can help their work. Start with high-value, low-regret tasks: writing and synthesis, planning and analysis, knowledge search, and repetitive administrative work. Build a curated "AI toolkit" of prompts, templates, and workflows. Then evolve toward agents that execute autonomously within your existing systems.

Third, embrace continuous learning. AI capabilities are shifting weekly. The advantage isn't "knowing AI," but learning faster than AI evolves. Focus on fundamentals—how models fail, how to evaluate output quality, how to protect data, and how to design human-in-the-loop controls—so you can adapt as tools evolve.

The Reality Check and Path Forward

AI is moving fast, but it is not magic. Many organizations will remain stuck in pilots until they redesign processes, strengthen data foundations, and build measurable governance. Strategy with execution discipline at scale is key. Enterprises need to build architecture and operating models that can absorb new capabilities without destabilizing the enterprise. The winners will be those who industrialize AI grounded in trusted data, redesigned workflows, accountable governance, and measurable outcomes.

As for what's next? One emerging frontier worth watching is spatial intelligence: World-model approaches aim to help AI perceive and reason about 3D environments, so systems can interact with the physical world more reliably.[19] It will be exciting to see how this plays out. At the end of the day, the strategic imperatives will still apply as will our need to continually embrace the ever-evolving workplace of the future.

About the Author

Irene Yang is a Product Innovation and Digital/AI Transformation Leader and founder of Finesse Innovation in New York. As a former

AWS Principal in innovation, corporate innovation director (Grant Thornton), management consultant (EY, Accenture), and VP of Private Equity Portfolio Operations, she helps leadership teams turn digital and AI initiatives into value creation – improving speed-to-market, cost-to-serve, and decision quality through disciplined execution. Irene leads work from diagnostic and discovery through build and scale, modernizing operating models and delivering product platforms and ML/AI-enabled capabilities. She operationalizes adoption with product instrumentation, KPI cadence, and governance so impact is measurable, repeatable, and sustainable.

Email: irene@finesseinnovation.com

References

1. Precedence Research, *Artificial Intelligence (AI) Market Size, Share and Trends, 2025–2034*, September 29, 2025, https://www.precedenceresearch.com/insights/artificial-intelligence.

2. Precedence Research, *Generative AI Market Size, Share and Trends, 2025–2034*, May 22, 2025, https://www.precedenceresearch.com/generative-ai-market.

3. McKinsey & Company, *The State of AI in 2025: Agents, Innovation, and Transformation*, November 2025, https://www.mckinsey.com/capabilities/quantumblack/our-insights/the-state-of-ai.

4. Gallup Workplace, "AI Use at Work Has Nearly Doubled in Two Years," June 15, 2025, https://www.gallup.com/workplace/691643/work-nearly-doubled-two-years.aspx.

5. OpenAI, *The State of Enterprise AI: 2025 Report*, December 8, 2025, https://openai.com/index/the-state-of-enterprise-ai-2025-report/.

6. Wharton Human-AI Research and GBK Collective, *Accountable Acceleration: Gen AI Fast-Tracks into the Enterprise*, October 27, 2025, https://knowledge.wharton.upenn.edu/special-report/2025-ai-adoption-report/.

7. D. Edge et al., "From Local to Global: A Graph RAG Approach to Query-Focused Summarization," *arXiv* preprint arXiv:2404.16130, April 24, 2024, https://arxiv.org/abs/2404.16130.

8. Microsoft Research, "Project GraphRAG," February 13, 2024, https://www.microsoft.com/en-us/research/project/graphrag/.

9. Reuters, "Over 40% of Agentic AI Projects Will Be Scrapped by 2027, Gartner Says," June 25, 2025, https://www.reuters.com/business/over-40-agentic-ai-projects-will-be-scrapped-by-2027-gartner-says-2025-06-25/.

10. Gartner, "Gartner Identifies the Top Strategic Technology Trends for 2026," press release, October 20, 2025, https://www.gartner.com/en/newsroom/press-releases/2025-10-20-gartner-identifies-the-top-strategic-technology-trends-for-2026.

11. NVIDIA Research, "Small Language Models Are the Future of Agentic AI," 2025, https://research.nvidia.com/labs/lpr/slm-agents/.

12. NVIDIA Developer Blog, "Traditional RAG vs. Agentic RAG: Why AI Agents Need Dynamic Knowledge to Get Smarter," July 21, 2025, https://developer.nvidia.com/blog/traditional-rag-vs-agentic-rag-why-ai-agents-need-dynamic-knowledge-to-get-smarter/.

13. The Linux Foundation, "Linux Foundation Announces the Formation of the Agentic AI Foundation (AAIF)," December 9, 2025, https://www.linuxfoundation.org/press/linux-foundation-announces-the-formation-of-the-agentic-ai-foundation.

14. Anthropic, "Donating the Model Context Protocol and Establishing the Agentic AI Foundation," December 2025, https://www.anthropic.com/news/donating-the-model-context-protocol-and-establishing-of-the-agentic-ai-foundation.

15. Microsoft Tech Community, "Announcing Microsoft Marketplace Expands with New AI Apps and Agents," 2025, https://techcommunity.microsoft.com/blog/microsoftpartnercommunity-blog/announcing-microsoft-marketplace-expands-new-ai-apps-and-agents/4400363.

16. Amazon Web Services, "AWS Summit New York 2025: Building AI Agents and a $100 Million Investment to Accelerate Agentic AI," June 25, 2025, https://www.aboutamazon.com/news/aws/aws-summit-new-york-2025-building-ai-agents.

17. IBM Institute for Business Value, *2025 CDO Study: The AI Multiplier Effect*, November 13, 2025, https://www.ibm.com/thought-leadership/institute-business-value/en-us/report/2025-cdo.

18. PwC, *PwC's 2025 Responsible AI Survey: From Policy to Practice*, October 30, 2025, https://www.pwc.com/us/en/tech-effect/ai-analytics/responsible-ai-survey.html.

19. World Labs, "Marble: A Foundation Model for Spatial Intelligence," November 10, 2025, https://www.worldlabs.ai/blog/marble-a-foundation-model-for-spatial-intelligence.

20. McKinsey & Company, "The agentic organization: Contours of the next paradigm for the AI era", September 2025. https://www.mckinsey.com/capabilities/people-and-organizational-performance/our-insights/the-agentic-organization-contours-of-the-next-paradigm-for-the-ai-era

CHAPTER 34

THE CULTURE OF TRUST: LEADING AI ADOPTION BEYOND POLICY AND CODE

By Joe Zhou
Founder of Complyd, AI Compliance
Sydney, Australia

The best way to find out if you can trust somebody is to trust them.

—Ernest Hemingway

The Trust Gap

After years spent leading AI teams and assisting companies in their preparation for complex enterprise deals, I have come to a singular realization: The technology itself is rarely the primary obstacle. It is a scenario I have witnessed time and again: Companies possess brilliant, sophisticated AI capabilities yet struggle profoundly to get their workforce to actually utilize them.

The metrics often paint a confusing picture. I have seen teams that boast exceptional customer satisfaction scores, sometimes reaching as high as 85%, yet their internal engagement with these expensive models sits as low as 10%. You end up with state-of-the-art models sitting idle, consuming resources, while talented people continue to execute workflows manually, just as they always have.

This disconnect extends beyond internal adoption. I have watched promising startups lose massive enterprise contracts, not because their artificial intelligence failed to perform, but because they lacked the ability to demonstrate that they were using it responsibly. The true barrier in these situations is rarely capability; it is trust.

An organization can possess the most comprehensive AI policy in the world, maintain the cleanest codebases, and deploy the most sophisticated models available. However, these technical assets are rendered effectively useless if the people meant to use them lack trust in the AI, doubt leadership's intentions, or do not feel safe enough to experiment and fail. Furthermore, as the market matures, if your customers and enterprise buyers cannot trust your AI governance, you will likely find yourself disqualified before you ever get the chance to prove that your technology works.

This chapter addresses the messy, often overlooked human side of AI adoption. It explores how to build a culture where trust is treated not as a corporate buzzword, but as the fundamental infrastructure that makes every other part of your strategy possible.

Part 1: Trust Starts at the Top

If you are a leader who expects your team to embrace AI, you must first integrate it into your own work. In my own experience, I have made a concerted effort to weave AI into my daily writing, planning, and coding tasks. The results have been tangible, leading to a five-time productivity boost in specific areas. For instance, drafting technical documentation, a task that previously consumed half my day, now takes approximately an hour. I rely on it to generate test scenarios for code reviews and to build the initial drafts of complex client proposals.

It is important to clarify that these are not theoretical numbers manufactured for a pitch deck; they represent actual, measurable improvements in how I handle my daily workload. However, the productivity gain itself is less important than what I do with that information. If I keep those wins to myself, they do little to change the culture.

The dynamic within my team shifted the moment I started sharing the reality of how I use AI, including the failures. I began discussing the prompts that led nowhere and the moments I had to abandon an AI-generated solution to return to manual work. This transparency prompted a change in the conversation. People stopped asking if we should use AI and started asking how others were successfully utilizing it.

I observed this pattern repeatedly during my four years as founding CTO at The Martec. As we prepared for our Series A funding, the temptation was to delegate the heavy lifting of AI integration to the team. Instead, I dedicated weeks to personally using our AI features for my own workflows.

When a feature failed to perform, I documented the issue publicly in our internal Slack. Conversely, when something worked brilliantly, I shared the exact prompting strategy I used. This stands in stark contrast to leaders who mandate AI adoption while remaining safely ensconced in their traditional workflows. These leaders speak of transformation while their assistants continue to format documents manually, or they push for "AI-first" development while approving code reviews the old-fashioned way.

Your team is always watching, and they notice this disconnect immediately. Such hypocrisy destroys trust far faster than any failed technical initiative ever could. While trust ultimately flows downward through an organization, it must originate at the top.

Part 2: Building Psychological Safety

Reflecting on my time leading engineering teams, the most difficult conversation I ever had wasn't about a system outage or a bug. It was about silence. Specifically, it was about why my team felt unable to speak up about their concerns regarding our AI features.

We had spent months building new AI capabilities, investing significant capital and resources, and launching with high confidence. The initial feedback was polite, and usage was acceptable, yet something felt off. It took weeks of probing in one-on-one meetings before someone finally leveled with me: The team didn't think the AI recommendations were good enough, but they hadn't wanted to slow down the roadmap.

That admission was a heavy blow. The pain didn't come from them being wrong about the AI, in fact, they were absolutely right, but from the realization that I had fostered an environment where people felt they couldn't voice quality concerns.

To fix this, we had to create explicit space for skepticism. In our AI adoption workshops, I stopped opening with a speech about why AI is amazing. Instead, I began by asking, "What are you worried about?" I now actively inquire about verification processes, potential failure modes, and the often large gap between polished AI demos and real-world reliability.

We also moved away from mandates in favor of "office hours," creating weekly sessions where developers share actual use cases and challenges without the pressure of performance reviews. In these sessions, one person might demonstrate how they used AI to write test cases, while another discusses their struggle to get decent API documentation suggestions. Because there is no pressure to prove ROI or hit arbitrary adoption numbers, the conversation remains honest.

Crucially, when someone experiments and fails, I ensure the entire team knows about it. This isn't to shame them, but to validate the act of experimentation itself. I might say, "Thanks for trying that approach and documenting why it didn't work. You just saved everyone else three hours." This signals that the value lies in the learning process, not just the outcome, and that we are exploring this territory together rather than performing for metrics.

Part 3: Transparency over perfection

One of the most effective steps I took for AI adoption was simply admitting what our AI couldn't do.

In the early days, we would demo AI features to customers using carefully crafted scenarios where the system performed beautifully. However, real-world usage was inevitably messier. Results were mixed, some were brilliant, others mediocre, and some were flat-out wrong. Customers remained polite, often saying things like "interesting" or "We'll evaluate it," but the subtext was clear: The product didn't match the promise, and, therefore, they didn't trust us.

We decided to change our approach by showing the full picture. We started presenting the good examples alongside the bad ones. We would say, "Here is a case where the AI nailed it. And here is one where it completely missed the mark, here is why that happened, and here is what we are doing about it".

Customer trust increased almost immediately. The technology hadn't changed, but our relationship with the customer had. We stopped overselling and started being honest.

This philosophy extends to how I communicate about my own AI-assisted work. Whenever I share content generated with AI support, I include context regarding how it was created and verified. A simple sentence like "I used AI to draft this analysis, then validated the data points against our actual metrics" does more for trust than any claim of high quality ever could. Perfection is unattainable with current AI, but transparency is a choice we can always make.

Part 4: Distributed Ownership

A significant mistake I made early in our adoption journey was assuming that the initiative needed to be driven entirely from the top. I had the vision, the strategy, and the resources, yet I lacked buy-in because the team felt that AI was something being done to them rather than with them.

The turning point came when I stopped mandating and started asking. We created a Slack Canvas with a simple, open-ended prompt: "Share one specific example of how you've used AI to enhance your productivity, creativity, or workflow." There were no quotas and no pressure, just genuine curiosity.

The responses were surprising and illuminated use cases I had never considered. Our customer success team was using AI to draft responses and then heavily editing them for tone. One engineer was generating test data, while another was using the tools to explain legacy code from an inherited system. None of these use cases originated in my strategy document; they came from people solving real problems in their daily work.

Previously, I thought we needed designated "AI champions" with specific roles responsible for driving adoption. However, I found that this often created a sense of obligation rather than enthusiasm. Now, I look for the people who are naturally experimenting and I amplify their work. These organic champions are far more effective because their advocacy is rooted in genuine experience rather than a job description.

When people feel ownership over how AI is utilized in their specific context, they become invested in its success. Conversely, when they feel they are simply executing someone else's vision, they inevitably become resistant.

Part 5: Measuring What Actually Matters

I must caution you against the worst metric I ever tracked: "AI feature usage." On paper, everything looked perfect. We had set adoption targets, usage was climbing, and the dashboard showed healthy engagement. In reality, people were using the AI features only because they were embedded in required workflows, not because they found them valuable. They would click the AI button, glance at the suggestions, ignore them, and proceed to do the work manually. We had achieved usage without achieving adoption.

True adoption manifests as changed behavior. It looks like a task that used to take three hours now taking 45 minutes. It looks like quality improvements that team members can articulate or teams proactively asking for AI capabilities in areas we haven't yet built them.

We found that the most valuable metrics were qualitative, gathered during our weekly office hours or customer workshops. When people started making comments like "I don't know how I worked without this" rather than a tepid "Yeah, it's useful sometimes," we knew we had achieved real adoption.

The ultimate test is simple: If you removed the AI features tomorrow, would anyone complain? Would they demand you bring them back? If the answer is "maybe" or "not really," you do not have adoption; you have compliance.

Part 6: When Trust Breaks

I want to share a story about the time I completely mishandled an AI rollout. We had been working on an AI-powered recommendation feature for months. The technology was solid, and the demos were impressive. Confident in our work, we shipped it to all customers simultaneously. Within 48 hours, the complaints began to roll in.

While the recommendations were technically correct, they were contextually tone-deaf. We had optimized for accuracy during testing but missed the nuanced context that mattered in real-world usage. The trust we had built over months began to erode in a matter of days.

My first instinct was to immediately dive into the code and fix the model. However, I realized that what customers needed first wasn't a technical solution, it was acknowledgment. We sent a message within 24 hours simply stating: "We've heard your feedback. You're right, this isn't meeting the quality bar you expect from us. Here's what we're seeing, and here's what we're doing about it".

That acknowledgment bought us the time and goodwill necessary to fix the issue properly. We created a public status page, updated every two days with our learnings and fixes. While some worried this transparency made us look incompetent, I believed it made us look honest. Our customers agreed; the feedback shifted from anger about the broken AI to appreciation for our transparency regarding the challenges.

This failure taught me that speed is not the ultimate goal; trust is. Now, when we launch AI features, we do so in stages: a small group of champions first, followed by learning and adjustment, then a slightly larger group, and finally a broader rollout. It is a slower process, but the resulting adoption is real and the trust remains intact.

I am still living with the consequences of that failed rollout. Some customers still mention it when we discuss new AI features, asking if the new tool will be "like the recommendations issue." That is the lingering cost of breaking trust. Trust is earned slowly and lost quickly, and one poorly executed rollout can undo months of work. The good news is that trust can be rebuilt, but it requires genuine change, radical transparency, and the humility to admit when you have moved faster than you should have.

Part 7: Trust Enables Compliance

It took me years to understand that trust and compliance are not competing priorities. They are, in fact, the same concept viewed from different angles. When people trust the AI and trust leadership's approach, compliance becomes significantly easier. Teams begin to proactively document decisions and raise risks early. They follow processes because they understand why those processes matter, rather than because they feel audited. I see this constantly in my work helping companies prepare for ISO 27001 and ISO 42001 certification and GDPR and EU AI Act compliance. Companies that treat compliance as a matter of cultural alignment rather than a checkbox exercise consistently achieve better adoption and better audit outcomes.

There is an uncomfortable truth that most companies discover too late: The questions coming from procurement are changing. Enterprise buyers used to simply ask, "Do you use AI?" Now, they ask for detailed controls regarding generative AI, specifically concerning data sovereignty, human oversight, and compliance with regulatory and IP obligations.

Most companies respond with a generic, "We use it responsibly." That answer is now an immediate red flag.

Consider this scenario: Your team uses Claude for client proposals. Can you document exactly which projects used it? Can you specify what client data went into the prompts, how the outputs were reviewed, and where that data was processed? Most organizations cannot answer these questions. However, in a trust-based culture, they wouldn't have to scramble to find the answers, the documentation would already exist.

In low-trust environments, compliance becomes adversarial. People hide problems, and documentation becomes retroactive fiction. Audits become stressful events where everyone scrambles to prove they did things they should have been doing all along.

In high-trust environments, compliance happens naturally. The psychological safety that allows employees to speak up about AI concerns is the same safety that makes them flag risks early. The transparency that builds customer trust is the same transparency that creates reliable audit trails. The distributed ownership that drives adoption is the same ownership that ensures processes are followed.

As AI governance becomes table stakes, two types of organizations will emerge: those scrambling to document what they should have been tracking and those walking into procurement conversations with confidence because trust was baked into their culture from the start. If you build the trust culture first, compliance becomes a byproduct rather than a burden.

The Trust Orbit

Trust is the foundation that enables proactive compliance and fearless adoption

Trust as Competitive Advantage

The companies that ultimately win with AI won't necessarily be the ones with the best models or the most aggressive adoption mandates. They will be the ones where people actually trust the AI, and perhaps more importantly, trust each other.

Trust compounds in ways that are difficult to see quarter over quarter but impossible to miss over the span of years. Early adopters share experiences, making it safer for others to try, while small successes build the confidence required for bigger experiments. Honest acknowledgment of failures makes people more willing to take necessary risks. And when the enterprise procurement team eventually shows up asking hard questions about your AI governance, you won't be scrambling, you will be ready.

I have shared my failures in this chapter as much as the successes because that is what trust requires: honesty about the messy reality, not just the polished success story. Every company struggling with AI adoption is dealing with some version of these challenges. The difference between those who succeed and those who don't is not that successful companies avoid these problems; it is that they build enough trust to work through them together.

If you are leading AI adoption, your job is not primarily technical. It is building trust. This means actually using AI yourself and sharing your process openly. It means creating space for people to question, doubt, and disagree. It means being radically transparent about what works and what doesn't, and giving people ownership over how AI shows up in their work. It means measuring what actually matters and responding to failures with honesty and genuine change.

Trust is the difference between AI adoption that looks good in reports and AI adoption that actually transforms how your organization works. Build the trust first. Everything else follows.

About the Author

Joe Zhou is the founder of Complyd.ai, building trust infrastructure for AI systems. He is an ISO/IEC 42001 lead auditor practitioner and EU AI Act specialist who helps startups and scale-ups achieve audit-ready compliance to win enterprise deals. Previously, Joe served as founding CTO at The Martec, where he built and scaled a high-trust engineering culture to serve global Fortune 500 clients and helped secure Series A funding while implementing ISO compliance frameworks. With over 18 years of experience in technology leadership, Joe specializes in the intersection of AI governance, enterprise readiness, and organizational trust. He is based in Australia and mentors founders in the AI space.

Email: hello@complyd.ai

Website: www.complyd.ai

DID YOU ENJOY THIS BOOK?

If you enjoyed reading this book, you can help by suggesting it to someone else you think might like it, and **please leave a positive review** wherever you purchased it. This does a lot in helping others find the book. We thank you in advance for taking a few moments to do this.

THANK YOU

You might also like other Thin Leaf Press titles:

The AI Advantage: Thriving Within Civilization's Next Big Disruption

The AI Revolution: Thriving Within Civilization's Next Big Disruption

The AI Mindset: Thriving Within Civilization's Next Big Disruption

AI: Work Smarter and Live Better Within Civilization's Next Big Disruption

Peak Performance: Mindset Tools for Managers

Peak Performance: Mindset Tools for Sales

Peak Performance: Mindset Tools for Leaders

Peak Performance: Mindset Tools for Business

Peak Performance: Mindset Tools for Entrepreneurs

Peak Performance: Mindset Tools for Athletes

The Successful Mind: Tools to Living a Purposeful, Productive, and Happy Life

The Successful Body: Using Fitness, Nutrition, and Mindset to Live Better

The Successful Spirit: Top Performers Share Secrets to a Winning Mindset

Winning Mindset: Elite Strategies for Peak Performance

Winner's Mindset: Peak Performance Strategies for Success

The Life Coach's Tool Kit, Vol. 1

The Life Coach's Tool Kit, Vol. 2

The Life Coach's Tool Kit, Vol. 3

Ordinary to Extraordinary

The Magical Lightness of Being

Explore.

www.ingramcontent.com/pod-product-compliance
Lightning Source LLC
Chambersburg PA
CBHW061257030726
47595CB00001B/103